AGRICULTURAL RESEARCH UPDATES

VOLUME 27

AGRICULTURAL RESEARCH UPDATES

Additional books and e-books in this series can be found on Nova's website under the Series tab.

AGRICULTURAL RESEARCH UPDATES

AGRICULTURAL RESEARCH UPDATES

VOLUME 27

PRATHAMESH GORAWALA
AND
SRUSHTI MANDHATRI
EDITORS

nova
science publishers
New York

Library of Congress Cataloging-in-Publication Data

ISBN: 978-1-53615-916-5
ISSN: 2160-1739

Published by Nova Science Publishers, Inc. † New York

CONTENTS

PREFACE

The aim of the opening chapter of Agricultural Research Updates. Volume 27 is to discuss the role of phenolic compounds in the interactions of plants with abiotic stress, with special attention paid to their antioxidant properties.

The second chapter attempts to assess the sustainability level of different types of agro-ecosystems in Bulgaria. A holistic hierarchical framework for measuring the integral, economic, social and ecological sustainability of agro-ecosystems in Bulgaria is proposed.

Following this, a study is presented with the goal of identifying and evaluating social and economic sustainability indices in corn production.

Statistical regression models were derived to estimate summer precipitation in a single location derived using lineal backward regression techniques, and the different efficiency between the derived models was analyzed.

Next, an analytical Fourier series solution to the equation for heat transfer by conduction in a spherical shell with an internal stone consisting of insulating material is presented, and an internal heat source linearly reliant on temperature is considered as a model of conduction of heat in stone fruits.

The subsequent study analyses the effects of certain parameters on the baking properties of wheat in years with different weather conditions. The results indicate statistically significant correlation coefficients between protein content, wet gluten content, alveograph properties, exstensograph properties and bread loaf volume.

Lastly, a summary of recent knowledge is evaluated in order to contribute to a better understanding of how farming practices affect berry composition and consequently, the sensory characteristics of raspberries.

Chapter 1 - Due to rising public pressure in recent decades, alternatives for large-scale and industrial farming are being sought. Environmental and sustainability issues and the rising awareness of the link between the overuse of pesticides/fertilizers and negative health effects have been key factors for creating the integrated production approach, which encompasses environmentally friendly technologies. Moreover, the demand for organically grown products is constantly growing. The organic production model is a step towards further restriction of synthetic chemical use in plant production. Limited use of pesticides may boost the plant's investment into its own defense systems, which may result in a higher content of secondary compounds. Synthesis of secondary metabolites is a common plant response to any form of stress (biotic or abiotic), and their function is to help the plant overcome unfavorable conditions. Many compounds, especially phenolics, are considered beneficial for human health. Phenolic substances are often thought of when referring to preformed resistance compounds. They are important in the protection of the plant against herbivores and pathogens. They may also play an important role in plant–plant interactions, in which they function as allelochemicals with either positive or negative effects on the neighboring plants. The aim of this chapter is to discuss the role of phenolic compounds in the interactions of plants with abiotic stress with special attention to their antioxidant properties.

Chapter 2 - Ecosystem approach has been increasingly incorporated in the management and evaluation of sustainability levels. Despite the huge progress in the theory and practice of this new area, there is still no consensus on how to assess the sustainability of agro-ecosystems due to diverse understandings, approaches, methods, employed data, etc. In

Bulgaria there are practically no deep studies on the sustainability level of diverse agro-eco-systems.

This chapter tries to fill the blank and assesses the sustainability level of agro-ecosystems of different type in Bulgaria. Firstly, a holistic hierarchical framework for assessing integral, economic, social and ecological sustainability of agro-ecosystems in Bulgaria is suggested, including 17 principles, 35 criteria, and 46 indicators and reference values. After that, an assessment is made on overall sustainability and its aspects of large (agro)ecosystems in North-Central, South-Eastern, South-Central and South-Western geographic regions, and particular of main and specific types of agro-ecosystems in the country - mountainous, plain-mountainous, plain, riparian (Struma, Maritza, Yantra), southern Black Sea, mountainous area with natural constraints, non-mountainous area with natural constraints, protected areas and reserves, Western Thracian Plain, Middle Danube Plain, Dupnitsa and Sandansko-Petrich Valley, Sredna Gora Mountains and Western Rila Mountains. The assessment is based on first-hand information collected though in-depth interviews with the managers of "typical" farms in the respective agro-ecosystems.

The study has found out that there is a considerable differentiation in the level of integral sustainability in agricultural ecosystems of different types. Furthermore, there are substantial variations in the levels of economic, social and ecological sustainability of agro-ecosystems of different type, and the critical indicators enhancing or deterring the overall and the particular sustainability of individual agro-ecosystems. The results of the integral agricultural sustainability level of this study, based on the micro agro-ecosystem (farm) data, are similar to the previous assessment based on the aggregate sectoral (statistical, etc.) data. There are large differences in the impact of socio-economic, institutional, behavioral, international, natural, etc. factors and individual public policy instruments on the sustainability of farming enterprises of different types and agro-ecosystems.

Having in mind the importance of holistic assessments of this kind for improving agricultural sustainability, farm management and agricultural policies, they should be expended and their precision and representation increased.

Chapter 3 - Background: Economic and social sustainability dimensions can be used as one of the most important basic factors in policy planning and sustainable production. The complexity and diversity of social and economic indices on the sustainability of production requires considering these indices based on their role and function. This study aimed at identifying and evaluating social and economic sustainability indices in corn production.

Method: Delphi method and Analysis Network Process (ANP) were used to achieve this goal and determine the weight of each index used to establish the economic and social sustainability. The accumulation method had been also based on library studies and interviews with experts and officials.

Results: The research findings indicated that respectively, management dimension (0.467) and participation (0.392) have the maximum weight and as a result, have been the greatest effect between indices of economic and social sustainability.

Conclusion: In other words, the role of management decisions and planning compared to participation and policy-making factors should be highly considered among economic and social sustainability indices in corn production.

Chapter 4 - Northern Patagonia is a dry area with scarce precipitation and agricultural and livestock activities which are highly dependent on inter-annual rainfall variability. In this study statistical regression models were derived to estimate summer precipitation in a single location using two different sets of data: ERA - INTERIM and NCEP reanalysis. Both datasets were used to determine the predictors. The models were derived using lineal backward regression techniques and the different efficiency between the derived models was analyzed. Results pointed out the best resolution of ERA-INTERIM dataset, allowed the definition of more and better predictors with physical meaning and improved efficiency.

Chapter 5 - Analytical models for predicting postharvest cooling times at the stone–pulp interface in stone fruit, or for measuring the thermophysical parameters of the pulp, mostly consider the pulp as a passive object for purposes of heat conduction, without internal heat generation or with a constant heat generation assigned. However, this is not strictly true. Like all other fruits, vegetables, etc., stone fruits respire, generating a heat that increases with temperature and that should also be taken into account in equations. This chapter presents an analytical Fourier series solution to the equation for heat transfer by conduction in a spherical shell with an internal stone consisting of insulating material and considering an internal heat source linearly reliant on temperature, as a model of conduction of heat in stone fruits.

Chapter 6 - Extreme weather conditions have been occurring every year in certain regions of the world since the beginning of the 21st century; dry and rainy years alternate each other, and there are long periods with extremely high temperatures. Agricultural crop production, with its main task to feed the humankind, has definitely been affected by them. The effects of climate changes on crop yields are the general focal point of research studies, while crop quality is much less under the spotlight. This is especially true for wheat due to many parameters of its technological quality, the most important of which are the baking properties. The aim of this study was to analyse the effects of certain parameters of technological quality on the baking properties of wheat in years with different weather conditions. The results from this study showed that there were statistically significant correlation coefficients between protein content, wet gluten content, alveograph properties (work, extensibility and tenacity and extensibility ratio), exstensograph properties (energy, resistance and extensibility ratio) and bread loaf volume in the year with dry and hot weather from anthesis to full grain maturity. The same parameters were not correlated with bread loaf volume in the year with high precipitation from anthesis to full grain maturity.

Chapter 7 - Organic farming is distinguished from other production systems because it excludes almost all synthetic chemicals, uses cultural rotations, crop residues, animal manure, and all organic waste from the farm. Several studies have demonstrated that fruits and vegetables from organic farming are generally endowed with enhanced physicochemical and sensory properties. Red fruits should be part of a balanced diet due to their characteristic and specific content of phenolic compounds, consisting essentially of anthocyanins, flavonols, proanthocyanidins and phenolic acids, known for their high antioxidant capacity. However, the studies described by different authors are inconsistent when indicating that organic and conventional farming of fruits and vegetables differs concerning physicochemical and sensory characteristics. So, the aim of this chapter is to summarize recent knowledge in order to contribute to a better understanding of how farming practices affect berry composition and consequently, the sensory characteristics of raspberries.

In: Agricultural Research Updates. Volume 27 ISBN: 978-1-53615-916-5
Editors: Prathamesh Gorawala et al. © 2019 Nova Science Publishers, Inc.

Chapter 1

THE ROLE OF PLANT PHENOLICS IN INDUCTION OF PLANT DEFENSE SYSTEM

Heba I. Mohamed**, Maged M. Abou-El-Enain and Eman M. Fawzi
Faculty of Education, Biological and Geological Sciences Department,
Ain Shams University, Cairo, Egypt

ABSTRACT

Due to rising public pressure in recent decades, alternatives for large-scale and industrial farming are being sought. Environmental and sustainability issues and the rising awareness of the link between the overuse of pesticides/fertilizers and negative health effects have been key factors for creating the integrated production approach, which encompasses environmentally friendly technologies. Moreover, the demand for organically grown products is constantly growing. The organic production model is a step towards further restriction of synthetic chemical use in plant production. Limited use of pesticides may boost the plant's investment into its own defense systems, which may result in a higher content of secondary compounds. Synthesis of secondary

* Corresponding Author's Email: hebaibrahim79@gmail.com.

metabolites is a common plant response to any form of stress (biotic or abiotic), and their function is to help the plant overcome unfavorable conditions. Many compounds, especially phenolics, are considered beneficial for human health. Phenolic substances are often thought of when referring to preformed resistance compounds. They are important in the protection of the plant against herbivores and pathogens. They may also play an important role in plant–plant interactions, in which they function as allelochemicals with either positive or negative effects on the neighboring plants. The aim of this chapter is to discuss the role of phenolic compounds in the interactions of plants with abiotic stress with special attention to their antioxidant properties.

Keywords: plant defense, abiotic stress, antmicrobial, fungi, insects

INTRODUCTION

Plant defense mechanisms are activated in a very short time in response to harmful agents. One of the first defense reactions observed in the cells of all living organisms is the increase in the concentration of reactive oxygen species (ROS), characterized by a very high chemical reactivity (Bartosz, 2004). The free radicals include molecules such as superoxide anion ($O2^{\bullet-}$) and extremely reactive hydroxyl radical (OH•) (Kozłowska and Konieczny, 2003). One of the most common defense reactions in the cells of all living organisms is the rapid increase in concentration of reactive oxygen species called oxidative burst (Bolwell 1999). Due to the universality of this phenomenon, understanding the role of ROS in defense mechanisms is an important aspect of plant stress physiology research. The main place for the formation of ROS during plant stress responses are plant cell wall and membrane. It is believed that the production of ROS is a result of enzyme activity located within these structures. It has been experimentally proven that oxidative burst is absent in isolated protoplasts (plant cells without cell walls) (Kozłowska and Konieczny, 2003). Thus, the presence of the cell wall is necessary to initiate the oxidative burst. Generation of ROS within the cell wall and their release outside the cell appears to be intentional, allowing their direct toxic effect on the cells of pathogens, contributing to the reduction of infection (Nowicka and Kruk 2013).

Under biotic stress conditions, salinity, herbicides and presence of Cd^{2+} ions in soil, large amount of $O2^{\bullet-}$ are formed in the peroxisomes which in physiological conditions are involved in H_2O_2 detoxification (Demidchik, 2015). The high concentration of reactive oxygen species is harmful to plant cells. Prolonged exposure to environmental stress factors results in so called 'oxidative stress' and leads to the pathological condition. ROS initiate free radical reactions that lead to oxidation of proteins, lipids and nucleic acids.

Degradation of essential cell components results in impairment of their function, eventually leading to death (Puzanowska-Tarasiewicz et al., 2008). The main source of the most toxic hydroxyl radicals is Fenton's reaction which takes place in cytoplasm. The reaction is catalyzed by transition metals including iron, copper, nickel, zinc, cobalt, manganese or chromium (Bartosz, 2004). Besides the destruction of biological membranes, ROS are also capable of reacting with proteins which leads to numerous modifications of amino acid residues or non-protein prosthetic groups. The most reactive hydroxyl radical may detach protons from protein molecules, resulting in the formation of protein radicals. Occurrence of non paired electrons on cysteine residues leads to the formation of disulfide bridges and, in a consequence, creation of protein dimers. The activity of ROS may lead to fragmentation or aggregation of proteins and, therefore, modified proteins lose their function (Bartosz, 2004). Nucleic acids, as a genetic information carriers are more resistant to destabilizing activity of ROS. The indirect oxidative DNA and RNA damage might result only due to hydroxyl radical (Bartosz, 2004). In order to protect their own tissues, plants have developed an efficient antioxidant system. It includes both antioxidant enzymes, which catalyze the decomposition of ROS, as well as non-enzymatic antioxidants including well known ascorbic acid (vitamin C), tocopherols and tocotrienols (vitamin E), carotenoids, numerous of phenolic compounds, as well as tripeptide glutathione (γ-L-Glutamyl-L-cysteinyl-glycine). Maintaining high antioxidant activity allows plants to efficiently capture toxic ROS and determines their tolerance to stress (Gill and Tuteja, 2010).

Phenolic compounds are a highly diverse class of secondary metabolites that are widely distributed among plants (Martens et al., 2010). Different physiological roles have been attributed to phenolics, including the protection of plants from abiotic and biotic stresses (Dixon, 2001; Treutter, 2005; Agati et al., 2012).

Phenolic compounds are synthesized from aromatic amino acids, predominantly from phenylalanine, and therefore are characterized as having at least one aromatic ring with one or more hydroxyl groups attached (Treutter, 2010). Phenolics range from simple, low molecular-weight compounds with a single aromatic ring, such as phenolic acids, to large and complex molecules such as tannins. The most prominent group of phenolics is the flavonoids with two aromatic rings. Sugars, organic acids, and other compounds are often bound to the rings, and the compounds are water-soluble. However, some flavonoids form large water-insoluble polymers. They are mostly stored in vacuoles of the epidermal cell of plant organs or specialized cells fulfilling their function as secondary metabolites. Another important group of phenolics, chemically related to flavonoids, are stilbenes with resveratrol as the most preminent compound in this group (Bavaresco et al., 2012). Similar to other secondary metabolites, phenolics have been linked to numerous ecological functions. They often act as a part of plant responses to biotic and abiotic stimuli. They are not only indicators of plant stress responses to various abiotic factors such as intensive light or mineral deficiencies, but are also one of the most important compounds in plant resistance and tolerance against pests and diseases. They represent a biochemical basis for successful reproduction of the plants, which synthesize them to attract pollinators or seed-dispersing animals. Phenolic-based polymers, such as lignin, suberin, or tannins, contribute substantially to the stability and robustness of plant tissues, towards mechanical or environmental damage, such as drought or wounding (Treutter, 2010). Phenolics can act as allelopathic compounds. A good example is juglone, a chemical released from the decomposing plant material of *Juglans nigra* (black walnut).

Plants which synthesize and release allelochemicals mostly influence the performance of neighboring plants in a negative way, gaining advantage in nutrient, water, or light acquisition. Targeted positioning of allelopathic plants could be utilized in crop rotations with the aim of reducing pesticide use and enhancing crop yields (Bavaresco et al., 2012).

Plants are constantly in contact with other organisms including those that are pathogenic which try to get nutrients from plants. Plants have developed defense mechanisms to protect themselves, and pathogens have established countermeasures to overcome these mechanisms. Plants invest in physical (e.g., cell wall, cuticle, trichomes) and chemical barriers (different secondary compounds), which should protect them from invading organisms. These barriers can be either pre-formed stored phytochemicals (phytoanticipins) or induced with pathogen attack (*ed novo* synthesized phytoalexins) (Piasecka et al., 2015).

The Role of Phenolic Compounds in Plant Defense against Biotic Stress

Phenolics serve a dual function of both repelling and attracting different organisms in the plant's surroundings (Table 1). They act as protective agents, inhibitors, natural animal toxicants and pesticides against invading organisms, i.e., herbivores, nematodes, phytophagous insects, and fungal and bacterial pathogens (Dakora and Phillips, 2002; Lattanzio et al., 2006). Simple phenolic acids, complex tannins and phenolic resins on the plant surface deter birds by interacting with the gut microflora and diminishing their digestive ability. The scent and pigmentation conferred by low-molecular-weight phenylpropanol derivatives attract symbiotic microbes, pollinators and animals that disperse fruits (Vit et al., 1997; Ndakidemi and Dakora, 2003).

Table 1. Examples of different types of phenolics and diverse functions that they perform for plants and bacteria

Examples of Phenolics	Functions
Coniferyl alcohol, sinapinic acid, cinnamic acid	vir gene inducers, determinants of scent and attractants of pollinators and symbiotic microbes in plants, etc
Hydroxybenzoate, hydroxycinnamates, 5-hydroxyanthraquinones Umbelliferone, p-hydroxybenzoic acid, vanillyl alcohol, isoflavones Sinapinic acid, syringic acid, ethylsyringamide, propylsyringamide, carbethoxyethylensyringamide, parahydroxybenzoate, ferulic acid Vanillyl alcohol, bromo acetosyringone	Allelochemicals for plant competition Chemoattractants in Rhizobium vir gene inducers in Agrobacterium Inhibitors of vir gene induction in Agrobacterium
Acetosyringone, a-hydroxyacetosyringone, p-hydroxybenzoate	Chemoattractants in Agrobacterium and Rhizobium, and vir gene inducers in Agrobacterium
Salicylic acid	Quorum quencher in Agrobacterium
Hydroquinones	Allelochemical for plant competition
Coumarins, xanthones, anthocyanidins	Determinants of colour and attractants of pollinators in plants
Caffeic acid	vir gene inducer in Agrobacterium
3,4-Dihydroxybenzoic acid	Chemoattractant in Agrobacterium and Rhizobium
Protocatechuic acid, b-resorcylic acid, protocatechuate, p-resorcylate, catechol	vir gene inducer in Agrobacterium
Chlorogenic acid	Precursor for lignin and suberin synthesis in plants
Lignin, tannins and suberins	Structural components of plant cell
Catechins	Plant defence
Flavonoids, flavonols, flavones, genistein, daidzein, O-acetyldaidzein, 6-O-malonylgenistin, 6-O-malonyl daidzin, glycitin, 6-O-malonylglycitin	nod gene inducers in Rhizobium Apigenin, naringenin, luteolin
Gallate, gallic acid, pyrogallic acid, syringic acid, kaempferol	vir gene inducers in Agrobacterium
Flavanones, quercetin	nod gene inducers in Rhizobium
Cajanin, medicarpin, glyceoline, rotenone, coumestrol, phaseolin, phaseolinin, limonoids, tannins, flavonoids defence	Phytoalexins, phytoanticipins and nematicides in plant

Phenolics have long been recognized as phyto-estrogens in animals (Adams, 1989) and as allelochemicals for competitive plants and weeds (Weir et al., 2004; Xuan et al., 2005). Mainly, the volatile terpenoids, the toxic water-soluble hydroquinones, hydroxybenzoates, hydroxycinnamates and the 5-hydroxynapthoquinones are widely effective allelochemicals. A number of simple and complex phenolics accumulate in plant tissues and act as phytoalexins, phytoanticipins and nematicides against soil-borne pathogens and phytophagous insects (Akhtar and Malik, 2000; Lattanzio et al., 2006). Therefore, phenolic compounds have been proposed for some time to serve as useful alternatives to the chemical control of pathogens of agricultural crops (Langcake et al., 1981). Most known effects of polyphenols on microbes are negative (Cushnie and Lamb, 2005; Ferrazzano et al., 2009). Plants respond to pathogen attack by accumulating phytoalexins, such as hydroxycoumarins and hydroxycinnamate conjugates (Mert-Türk, 2002; Karou et al., 2005). The synthesis, release and accumulation of phenolics in particular, salicylic acid (Koornneef and Pieterse, 2008; Lu, 2009) are central to many defense strategies employed by plants against microbial invaders.

The Role of Phenolic Compounds in Plant Rhizosphere

The plant rhizosphere is a dynamic ecosystem of different species, representing flora, fauna and microbes that interact with each other in a variety of complex reactions (Whipps, 2001). These interactions are mainly governed by a diverse array of phenolics exuded by the growing roots of plants, together with a host of other chemicals (Bertin et al., 2003; Bais et al., 2006). The root exudates generally include ions, free oxygen, water, enzymes, mucilage, a number of carbon-containing primary and secondary metabolites and, most importantly, plant phenolics (Dakora, 2003). The released phenolics differ from species to species and also with time, space and location. Their concentrations in the soil range from 2.1 to 4.4% in monocotyledonous plants and 0.1 to 0.6% in dicotyledonous plants (Hartley and Harris, 1981). Phenolics trigger redox reactions in soils and selectively influence the growth of soil microorganisms that colonize the rhizosphere. These then influence the hormonal balance, enzymatic activity, availability of phytonutrients and competition between neighboring plants (Hättenschwiler and Vitousek, 2000; Kraus et al., 2003). As a result of this dynamic and ever-changing interaction, the structure and chemistry of the soil are altered significantly depending on the quantity and identity of phenolics released by different plant species. The composition of the microorganism species in different root locations is also persistently modified and shaped. Moreover, as phenolics move through the rhizosphere, they are bound by soil organic matter and metabolized by the bacterial flora of the soil (Kefeli et al., 2003).

Microorganisms break down phenolics into elements that contribute towards the mineralization of soil nitrogen and the formation of humus (Halvorson et al., 2009). The phenolics chelate metals and improve soil porosity, providing active absorption sites and increasing the mobility and bioavailability of elements, such as potassium, calcium, magnesium, copper, zinc, manganese, molybdenum, iron and boron, for plant roots (Seneviratne and Jayasinghearachchi, 2003). Some phenolic metabolites, such as *trans*-cinnamic acid, salicylic acid, coumarin, benzoic acid, parahydroxybenzoic acid and syringic acid, are phytotoxic. For example, the accumulation of phenolics in the soil can inhibit seed germination and seedling growth (Baleroni et al., 2000). This effect may be caused by interference with cell division and the normal functioning of cellular enzymes. Indeed, phenolics have been shown to inhibit phosphatase and prolyl aminopeptidase involved in seed germination (Madhan et al., 2009). In addition, phenolics affect the process of mineral uptake by the plants (Lodhi et al., 1987). Many of the phenolic root exudates serve as chemotactic signals for a number of soil microorganisms that recognize them and move towards plant roots in the carbon-rich environment of the rhizosphere (Perret et al., 2000; Taylor and Grotewold, 2005). Based on the nature and type of root-derived chemicals, both positive and negative cross-talk pathways are initiated between roots, roots and insects, and roots and microbes. Different organisms are repelled from or attracted to the same chemical signal, which then elicits different responses in different recipients. One specific example is the isoflavones from soybean roots which serve as a chemoattractant for both the symbiotic *Bradyrhizobium japonicum* and the pathogenic *Phytopthora sojae* (Morris et al., 1998). This subsequently leads to the colonization of roots. Generally, the zone of root elongation, just behind the tip, supports the growth of primary root colonizers that utilize the easily degradable sugars and organic acids. On the other hand, fungi and bacteria that are adapted to crowded, oligotrophic conditions inhabit the older root zones, comprising sloughed cells with lignified cellulose, hemicellulose and carbon deposits. Mature communities of fungi colonize relatively nutrient-rich environments provided by the newly emerging lateral roots and the secondary non-growing root tips. Colonization of vesicular arbuscular mycorrhizal fungi in response to isoflavonoids from soybean roots leads to greater phosphorus acquisition for plant nutrition, improved water relations and, consequently, better plant growth (Bagayoko et al., 2000). During phosphate deficiency, plant roots also exude strigolactones. These apocarotenoid molecules are detected as host-derived signaling compounds at the presymbiotic stage of beneficial fungal symbionts (Akiyama et al., 2005; Akiyama, 2007). Unlike various flavonoids that induce hyphal branching only in a limited number of hosts, strigolactones represent primary signaling factors for hyphal branching and growth of arbuscular mycorrhizal fungi (Steinkellner et al., 2007).

Other positive effects of root colonizers include the symbiotic associations with epiphytes and mycorrhizal fungi, the fixation of atmospheric nitrogen by different classes of proteobacteria (Moulin et al., 2001), increased biotic and abiotic stress tolerance imparted by the presence of endophytic microbes (Schardl et al., 2004), and several direct and indirect advantages caused by various plant growth-promoting rhizobacteria (Gray and Smith, 2005). Some colonizing bacteria also interact with plants to produce protective biofilms or antibiotics, functioning as an effective biocontrol against potential pathogens (Bais et al., 2004).

The Role of Phenolic Compounds as Antimicrobial and Insecticidal Metabolites

When one considers resistance expression in plants, it is necessary to consider whether resistance expression is part of a passive or active response system. There are several situations that could arise:

1. Compound "A" is present in the plant and is toxic to the potential pathogen. The compound is present in cells or tissues that the pathogen must come into contact with, at some time during the attempted infection. The compound is not further metabolized, but may or may not be changed by the pathogen, and the compound as such then accounts for its toxicity to the pathogen. This example represents a form of passive resistance.
2. A preformed substance is degraded or metabolized to a different compound by the host in direct response to the pathogen and it is this compound that accounts in part for toxicity to the pathogen. Because the host changes the compound, this would be considered a mechanism of active resistance.

1. Passive Resistance

Phenolic substances are often thought of when referring to preformed resistance compounds. The following classes of metabolites, however, should also be included: alkaloids, carbohydrates that can bind bacteria, proteins that act as lectins, amino acids, terpenoids, and polyacetylenes. Some examples of these compounds are:

1.1. Chlorogenic Acid

Chlorogenic acid (1) is an example of a preformed compound that has a relatively low level of toxicity to most microorganisms. It is effective against microorganisms considered as weak pathogens of potato. In potato tubers, chlorogenic acid is present in the periderm (Kojima and Kondo, 1985) and is toxic to the organism that causes potato scab, *Streptomyces scabies*. In general, there is more of the compound in tubers of cultivars that are resistant to the pathogen than in tubers of cultivars that are susceptible. Chlorogenic acid also affects the growth of the vascular pathogen *Verticillium albo-atrum* and is present in the vascular tissue of the potato (Dao and Friedman, 1994). The mechanism by which *ortho*-dihydroxy phenolic compounds such as chlorogenic acid provide defense against insect pests was studied by Felton et al., (1989). They investigated the fate of chlorogenic acid in tomato (*Lycopersicon esculentum*) on the feeding behavior of the tomato fruit worm (*Heliothis zea*) and beet army worm (*Spodoptera exigua*). Upon feeding by the insects, polyphenol oxidases that are compartmentally separated from chlorogenic acid in the plant, come in contact with their substrate and convert chlorogenic acid to the toxic chlorogenoquinone (2). This quinone is a highly reactive electrophile and will react with nucleophilic –SH and –NH2 moieties in proteins (as indicated by structure 3) (Matheis and Whitaker, 1984). This results in the cross-linking of proteins with chlorogenic acid, which reduces the availability of free amino acids and proteins to the insect.

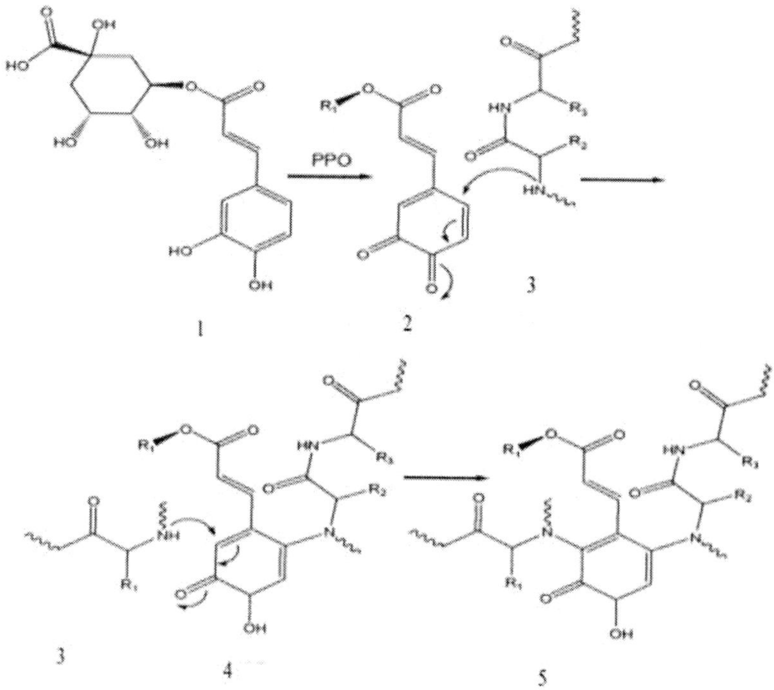

Figure 1. Oxidation of chlorogenic acid (1) by polyphenoloxidase (PPO), resulting in chlorogenoquinone (2), which can react with nucleophilic groups in proteins (3) to give the cross-linked compound 4, which can react with another protein molecule to yield 5.

1.2. Phloridzin and Phloretin

In apple, the glycoside phloridzin (Figure 2) and its aglycone phloretin are thought to inhibit the apple scab fungus *Venturia inaequalis*. Phloridzin is also an *ortho*-dihydroxyphenolic compound, and like chlorogenic acid can also be easily converted to a reactive quinone upon attack by a pathogen. Raa (1968) demonstrated that oxidation products of phloridzin inhibit fungal pectinases. Fungal pectinases hydrolyze pectin, a cell wall compound that is abundant in the middle lamella and plays a role in cell adhesion. Thus, by inhibiting pectinases, the ability of the fungus to hydrolyze and invade the plant cell wall would be compromised (Nicholson and Rahe, 2004). Also, Raa and Overeem (1968) have shown that phloridzin (XI) oxidation products may be involved in the defense mechanism of apple leaves against the scab fungus *Venturia inaequalis*.

1.3. Cyanogenic Glycosides

Many plants contain cyanogenic glycosides|. Toxicity of the cyanogenic glycoside results when the compound is enzymatically cleaved to release hydrogen cyanide (HCN) that is toxic to the pathogen. Sorghum contains the cyanogenic glycoside (Figure 2) dhurrin. This compound is of interest to both pathologists and entomologists as an example of a preformed resistance compound and acts as an insect feeding deterrent and as a fungicidal agent (Starr et al., 1984; Adewusi, 1990).

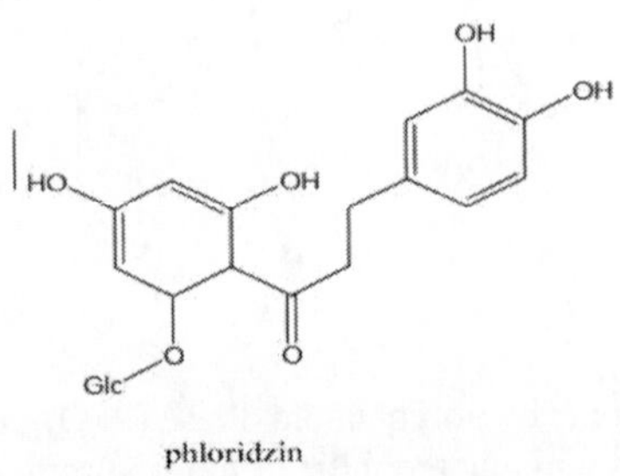

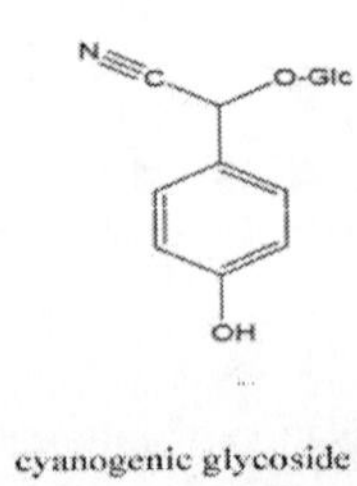

Figure 2. Example of phenolic compound.

1.4. Tuliposides

Tulips contain preformed compounds known as tuliposides, (tuliposide A, 1, and tuliposide B, 2). Hydrolysis of the tuliposides results in formation of aglycones (3) and (4) which will form butyrolactones (5) and (6). These lactone forms are inhibitory to fungi. Some fungi are able to metabolize the lactone form to a butyric acid form that is not inhibitory to a variety of organisms, including the tulip pathogen *Fusarium oxysporum*. *Botrytis tulipae* infects all parts of the tulip, including the pistils. *B. tulipae* converts tuliposides into hydroxycarboxylic acids which are non-toxic (Schönbeck and Schroeder, 1972).

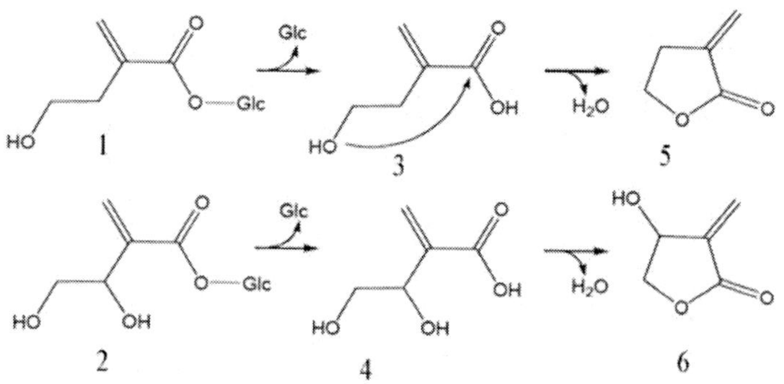

Figure 3. Formation of antifungal butyrolactones from tuliposides *via* an internal esterification reaction.

1.5. Protocatechuic Acid

Protocatechuic acid which is found in yellow and red skinned onions and prevents the germination of spores of the onion smudge fungus, *Colletotrichum circinans*. Onion scales accumulating sufficient quantities of catechol (I) and protocatechuic acid (II) help to prevent onion smudge disease, *Colletotrichum circinans*. The coloured outer onion scales of resistant onion varieties contain enough of these two phenols to reduce spore germination of *C. circinans* to below 2%, while susceptible varieties lack these compounds and the germination rate is over 90% (Anonymous, 1999).

1.6. Lignin

The cell wall polymer lignin has been implicated in defense against pests and pathogens, as a preformed, passive defense compound. It is a suitable defense compound in that respect, because it hardens the cell wall and thus creates a physical barrier against invasion. In addition, the chemical structure of lignin is very complex because of the different lignin subunits, and the many different types of chemical bonds that exist between subunits. Hydrolysis would, therefore, require a battery of enzymes, which is something most pests and pathogens do not have access to (Denton, 1998). Exceptions are white rot fungi, which can oxidize lignin in order to degrade it (Chen and Chang, 1985).

Deposition of lignin has been hypothesized to interfere with the enzymatic hydrolysis and mechanical penetration of plant tissue by fungal pathogens and may also impair the movement of water and diffusible molecules between the plant and fungus and then help to starve a pathogen. Lignin precursors themselves might exert a toxic effect on pathogens or, by binding to fungal cell walls, make them more rigid and impermeable, and so hindering further growth or uptake of water and nutrients. Thus, *in vitro* coniferyl alcohol is toxic to fungi at low concentration. Moreover, it has been proposed that the polymerization of lignin precursors by free radicals in the intercellular space might also lead to a lignification of pathogen structures: each of these possible events provides a testable hypothesis, and some of these mechanisms have been demonstrated (Friend 1985; Nicholson and Hammerschmidt, 1992; Sticher et al., 1997).

1.7. C-Glycosyl Flavones

Maysin (2”-*O*-α-l-rhamnosyl-6-C-(6-deoxy*xylo*-hexos-4-ulosyl)-luteolin; apimaysin and methoxymaysin are C-glycosyl flavones. These three compounds differ from each other in the substitution pattern of the B-ring. Maysin is two times more effective in its ability to inhibit growth of the corn earworm by the efficient formation of a toxic quinone, whereas the quinone formation from apimaysin and methoxymaysin is less efficient (Snook et al., 1994).

2. Active Resistance

Compounds formed in response to stress may occur in at least two ways. In one response, the plant may form compounds throughout the tissue at a considerable distance from the infection site (Hammerschmidt, 1999). In another response, the plant may form compounds specifically at the infection site (Snyder and Nicholson, 1990; Nicholson and Wood, 2001). In general, such compounds are referred to as either stress metabolites or more often as phytoalexins. By definition, phytoalexins are formed in response to infection (Hammerschmidt and Nicholson, 2001; Aguero et al., 2002). Phytoalexins often exhibit toxicity to specific pathogens.

2.1. 3-Deoxyanthocyanidins

The 3-deoxyanthocyanidins are a class of phytoalexins found in sorghum. These compounds are so fungi-toxic that they are effective at femtogram levels (Snyder and Nicholson, 1990; Nicholson and Wood, 2001). The synthesis of these compounds is initiated on the endoplasmic reticulum. Compounds are then trafficked in subcellular inclusions. The inclusions appear similar to vesicles, but there is no evidence that membranes surround the inclusions (Snyder and Nicholson, 1990). Nielsen et al. (2004) recently summarized this defense response. When the 3-deoxyanthocyanidins enter the apoplast, the host cell collapses. The phytoalexins then accumulate in the pathogen and cause its death. Excess phytoalexins are trapped in host cell walls at infection sites (Lo et al., 1998; 1999).

2.2. Pisatin

Pisatin is an isoflavonoid phytoalexin that is synthesized by pea (*Pisum sativum* L.) as a response to infection (Preisig et al., 1989). Subsequently, it was shown that pathogens capable of demethylating pisatin were tolerant of this phytoalexin. The enzyme responsible for demethylation is a specific cytochrome P450 mono-oxygenase released by the fungus *Nectria haematococca* (Delserone et al., 1999).

2.3. Salicylic Acid

When a specific part of a plant is attacked by a fungal pathogen, distant parts of the plant may display an enhanced state of resistance involving the accumulation of pathogenesis-related (PR) proteins, a class of plant proteins that are normally not present but that are induced upon pathogen attack (Van Loon and Van Strien, 1999). In addition, salicylic acid (SA) and hydrogen peroxide accumulate at the wound site and in other parts of the plant. This response is referred to as *systemic acquired resistance* (SAR) and is thought to be mediated by one or more signaling molecules in the phloem. SA was hypothesized to be one of those signaling molecules, because 1).

SA accumulation was shown to be correlated with SAR and resistance (Uknes et al., 1993), and 2) exogenous SA applied to an uninfected plant induced SAR and resistance in a manner similar to that of an infected plant (Ward et al., 1991), SA was hypothesized to be one of those signaling molecules, because 1) SA accumulation was shown to be correlated with

SAR and resistance (Uknes et al., 1993), and 2) exogenous SA applied to an uninfected plant induced SAR and resistance in a manner similar to that of an infected plant (Ward et al., 1991).

2.4. Lignin

There is evidence that lignin can be synthesized *de novo*. This lignin is synthesized locally, and specifically in response to pathogen attack. There is new evidence that this lignin requires different biosynthetic enzymes, which results in a different subunit composition than the lignin of the vascular tissue.

Wheat cultivars resistant to *Puccinia recondita* f. sp. *tritici*, a fungal pathogen causing leaf rust, were shown to accumulate more lignin than susceptible cultivars, based on histochemical stains and a quantitative assay to detect total phenolics (Southerton and Deverall, 1990). Several studies have focused on the role lignin biosynthetic enzymes play in response to pathogenic attack. Moerschbacher et al., (1988) used specific inhibitors of the enzymes phenylalanine ammonia lyase (PAL) and cinnamyl alcohol dehydrogenase (CAD). These inhibitors were applied to wheat cultivars highly resistant to stem rust (*Puccinia graminis* Pers f.sp. *tritici* Erics. & E. Henn.). Regardless of the inhibitor that was used, *de novo* lignification was decreased and fungal development increased. Thus, a strict correlation between resistance and lignification was demonstrated. Lignification in response to infection has been found to be associated by an increase activity of several enzymes of the lignin biosynthetic pathway. Deborah et al., (2001) studied PAL and peroxidase (PO) activity as well as accumulation of lignin in response to inoculation of rice leaf sheaths with a pathogen and a non-pathogen. Infection with the non-pathogen *Pestalotia palmarum* resulted in a stronger increase in PO and PAL activity and a higher accumulation of lignin than after inoculation with the pathogen *Rhizoctonia solani*. Altogether, these results indicate that if lignin plays a role in the resistance against a pathogen, a coordinated activation of lignin biosynthetic enzymes is required.

2.5. Tannins

An example of phenolic compounds protecting plants from herbivores is tannins. There are two types of these compounds: easily hydrolyzable tannins (produced by polymerization of gallic acid or other phenolic acids and some sugars) and condensed tannins (created by combining multiple flavonoid units). Tannins are commonly found in relatively high concentrations in the bark of trees and in leaves. Their repellent properties are due to their unpleasant, bitter taste, while the toxic effect is related to their ability to bind to and denature proteins. All these features make them an excellent compound providing plant protection against insects (Ashok and Upadhyaya, 2012).

Toxicity of tannins, hydrolysable tannins and proanthocyanidins, usually estimated by the measurement of the reduction of the *in vitro* growth of mycelium, is well documented for several filamentous fungi, for example *B. cinerea*, *Aspergillus niger*, *Colletotrichum graminicola*, *Gloeophyllum trabeum*, *Trichoderma viride*, and *Penicillium* sp.. Tannins are quite potent antibiotics. In temperate trees, tannins and related phenolic compounds in heartwood from fungal decay inhibit extracellular hydrolases from invading pathogens, thus preventing their rapid development in the plant (Harborne, 1980; Scalbert, 1991). It is possible that inhibition of extracellular fungal enzymes (cellulase, pectinase, laccase, xylanase, etc.), nutrient deprivation of substrates (metal complexation, protein insolubilization) and action on fungal membranes (inhibition of oxidative phosphorylation) are effectively involved in tannin toxicity.

2.5.1. The Role of Phenolic Compounds in Plant-Insect Interactions

Ehrlich and Raven (1964) have proposed that through the process of co-evolution, insects are able to detoxify certain defensive substances that deter feeding so that, eventually, the same deterrent compounds become feeding attractants. Finally, it must be emphasized that insects are reported to utilize secondary metabolites to increase their fitness (Figure 4). It is well known the ability of certain insects to sequester plant toxins from their food plants in the larval stage. They then move these toxins into the adult imago and both larva and adult generally gain protection from bird predation. Butterflies (*Polyommatus icarus*) reared at larval stage on inflorescences of *Vicia villosa*, a plant species acceptable for larval development even if not used as food source in nature, showed some important features of the uptake and sequestration of plant secondary metabolites by insects: (i) larvae incorporated only part of the flavonoids of their host plant (mainly quercetin and kaempferol, while other flavonoids were mostly excreted); (ii) the content of ingested flavonoids was correlated with the body mass; (iii) the flavonoids were metabolized and their conjugated forms were stored. These data also suggest that females are better adapted to sequester these compounds than males and that the accumulation of flavonoids in the smaller wings of the female is used in visual communication, as flavonoid-rich females are more attractive to males than flavonoid-free females (Burghardt et al., 1997, 2000, 2001).

Studies on the role of inhibitors in host plant selection indicate that many different chemicals may be expected to have an inhibitory effect on feeding by different insects. Among plant constituents, it is now generally accepted that plant phenolics play a role in protecting plants from both insect and mammalian herbivory (Corcuera, 1993; Harborne and Williams, 2000; Simmonds, 2003). Traditionally, methods for demonstrating that constitutive phenolics are involved in plant defense have depended on measuring the total phenolic content of plant tissues. While such measurements might occasionally show some correlation, experience shows that it is an individual subclass of phenolic or an individual structure that is active against a particular herbivore. In addition, one of the most important advances in understanding plant-insect interactions has been the discovery of the induced defence: some plants respond to insect feeding by increasing the synthesis of a particular phenolic toxin(s). Finally, the concentration of the toxic phenolic compound(s) in the plant is a key factor in deterrence and it is the accumulation of phenols in particular parts of the plant which represent a feeding barrier (Tosh et al., 2003; Zagrobelny et al., 2004).

A well-known example of simple phenolics being feeding barriers to insect herbivores is represented by salicylates in *Salix* leaves and their role in the feeding and growth of the polyphagous larvae of *Operophtera brumata*. It has been observed that the levels of salicylates correlated negatively with growth: larvae exposed to leaves rich in these compounds grew slowly and consumed less material, and this suggests that salicylates could be considered as antifeedants for *O. brumata* (Simmonds, 2003).

Chlorogenic acid, a phenylpropanoid derivative, showed antifeeding properties in the same system, even if, because of its widespread occurrence in the plant kingdom, this compound is less likely to be a useful defensive agent in that many insects may become adapted to it and hence would be expected to tolerate its dietary presence. The effects of pure chlorogenic acid on insect feeding behaviour were tested using four common leaf beetle species, which are in the field mainly found on willows with low-chlorogenic acid leaves. One species, *Lochmaea capreae* L., was invariably deterred by pure chlorogenic acid applied in naturally occurring concentrations on the willow leaves. Accordingly, in 2-choice laboratory feeding trials *L. capreae* was found to prefer low-chlorogenic acid leaves of four willow species over highchlorogenic acid leaves of *Salix pentandra* L. and *S. myrsinifolia* Salisb. When presented on the leaves of *S. phylicifolia* L, pure chlorogenic acid inhibited also the feeding by *Phratora polaris* Sp.-Schn. Instead, chlorogenic acid had no significant effect on *Ph. polaris* when it was presented on the leaves of another willow *S. cinerea* L. Thus, the response of *Ph. polaris* to chlorogenic acid seems to depend on the plant species. To two other leaf beetle species, *Galerucella lineola* F. and *Plagiodera versicolora* Laich., chlorogenic acid is an ineffective deterrent even at unnaturally high concentrations. Thus, among four studied leaf beetle species, only *L. capreae* seems to be clearly affected by this phenolic. Therefore, the overall importance of chlorogenic acid as a defence against willow-feeding leaf beetles appears to be very limited (Ikonen et al., 2001).

Plant flavonoids affect the behaviour, development and growth of a number of insects (Simmonds, 2001; Simmonds and Stevenson, 2001). Some flavonoids are feeding stimulants for the boll weevil, *Anthonomus grandis*, (Hedin et al., 1988), or oviposition stimulants of a Citrus-feeding swallowtail butterfly, *Papilio xuthus* L., (Nishida et al., 1987) or, finally, antibiotic substances efficient against phytophagous insects (Ridsdill-Smith et al., 1995).

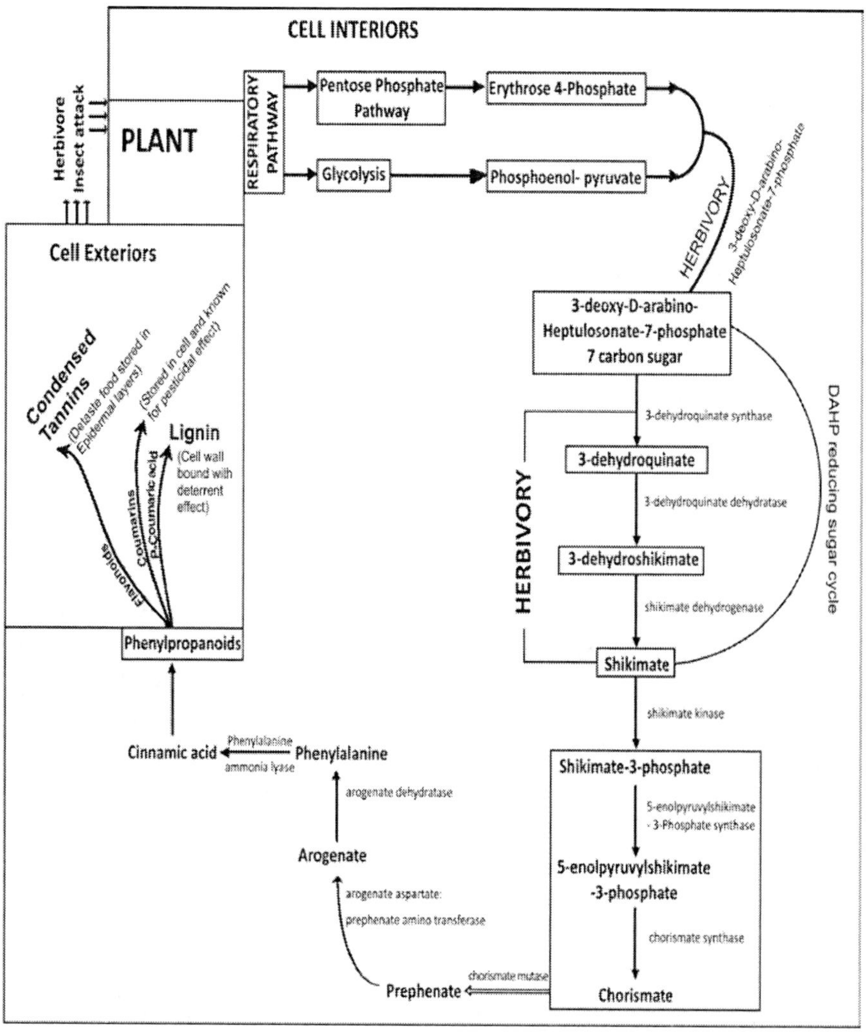

Figure 4. Biosynthesis, release and storage of defensive phenolics in plants attacked by herbivores. (Adapted and modified in parts from Razal et al., 1996).

Here it must be stressed that the salivary secretions of phytophagous insects contain various enzymes, playing a fundamental role in food digestion of sucking-piercing insects. Among these enzymes, polyphenol oxidases (PPO) [E.C. 1.10.3.1.] and peroxidases (POD) [E.C. 1.11.1.7], oxidoreductases metabolizing plant phenolics, are also present. Both enzymes have been identified in the salivary secretions of several aphid species and various functions have been proposed for these enzymes. At first, it was suggested that they are involved in the chemical stabilisation of the stylet sheath. Secondly, they were thought to enable phytophagous insects to overcome plant defenses by neutralising phenolics and their derivatives. In fact, it has been observed that when the stylets of grain aphids penetrate plant tissues, the ruptured cells may become partially filled with aphid saliva. Thus, the salivary sheath material may absorb plant metabolites from plant tissues. As a result, the grain aphid PPO secreted in its saliva can react with the phenolics accumulated around the aphid stylet sheath, thus oxidizing and polymerizing them. Natural occurring phenolic toxins are thus converted into less toxic substances (Urbanska et al., 1998).

It is also true that the effectiveness of phenolics as a resistance factor to insect feeding is enhanced by oxidation to polymers, which reduce digestibility, palatability and nutritional value. Thus high levels of PPO, the major phenolic oxidising enzyme of plants, can be correlated with plant resistance mechanisms against insects. Polyphenol oxidases, presumed antiherbivore enzymes, are antinutritive enzymes that decrease the nutritive value of wounded plant tissues by cross-linking proteins or catalyzing the oxidation of phenolic toxic metabolites to reactive and polymerizing quinones. The inducible expression of PPO was examined in hybrid poplar: following mechanical wounding, simulating insect damage, PPO activity increased in wounded and unwounded leaves on wounded plants beginning at 24h and 48h, respectively. Feeding by forest tent caterpillar also induced PPO expression. This wound- and herbivore-induced expression of PPO in hybrid poplar supports the defensive role of this enzyme against insect pests. The results obtained showed that: (i) PPO induction is systemic, i.e., induced PPO functions in reducing additional herbivore damage rather than in wound repair; (ii) PPO induction is mediated by increased mRNA accumulation, which is typical of a variety of induced defence proteins (Constabel et al., 2000; Haruta et al., 2001).

In addition, tannins are generally considered to be deleterious to phytophagous insects. Tannins may affect the growth of insects in three main ways: they have an astringent taste, which affects palatability, and decreases feed consumption, they form complexes of reduced digestibility with proteins and they act as enzyme inactivators. Tannins are present at high levels in most plant seeds and grains, where they negatively impact the use of seeds and grain in insect feed (Winkel-Shirley, 1998). In cowpea, condensed tannins (proanthocyanidins) contribute to resistance to infestation by insects such as the cowpea weevil *Callosobruchus maculatus* (F.) (Lattanzio et al., 2005). An increased level of proanthocyanidins during seed and grain development and maturation has been studied in a number of different plant species. A significant increase in the degree of polymerization of proanthocyanidins has been observed during maturation of sorghum grains (Butler, 1982). In *Phaseolus vulgaris* a gradual increase in condensed tannins and α- amylase inhibitory activity, both considered defensive tools against phytophagous insects, was observed from anthesis to seed maturity (Coelho and Majolo, 1993).

The mechanisms at pla are similar to those that enable them to resist defensive phytotoxins, and a functional overlap between adaptation to agrochemicals and to phytotoxins has been suggested by Dermauw et al., (2013) (Figure 5).

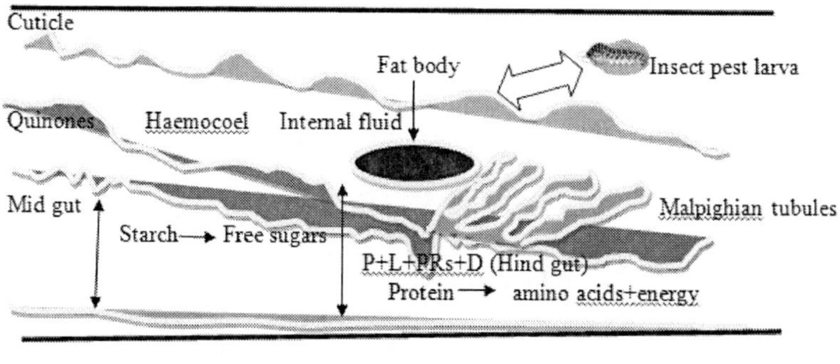

Figure 5. The response of herbivores towards the defensive secondary metabolites.

Conclusion

Phenolic compounds play important roles in plant growth and development, particularly in defense mechanisms. Most of the phenolic compounds have potent antioxidant properties. Importantly, phenolic phytoalexins exhibit antibiotic and antifungal activity. Coumarins and tannins repel herbivores, whereas phenylpropanoids are starting molecules for the synthesis of lignin and suberin, in order to strengthen cell walls.

Acknowledgments

Authors would like to show their appreciation to Faculty of Education, Ain Shams University, Department of Biological and Geological Sciences and to Professor Maged M. Abou-El-Enain, dean of Faculty of Education for their continuing cooperation to support the current research and provided facilities necessary to accomplish the most wanted objectives of the research.

References

Adams, N. R. (1989). Phytoestrogens. In: *Toxicants of Plant Origin* (Cheeke, P. R., ed.), pp. 23–51. Boca Raton: CRC Press.

Adewusi, S. R. A. (1990). Turnover of dhurrin in green sorghum seedlings. *Plant Physiol*ogy, 94:1219-1224.

Agati, G., Azzarello, E., Pollastri, S. and Tattini, M. (2012). Flavonoids as antioxidants in plants: location and functional significance. *Plant Science,* 196: 67-76.

Aguero, M. E., Gevens, A. and Nicholson, R. L. (2002). Interaction of *Cochliobolus heterostrophus* with phytoalexin inclusions in *Sorghum bicolor*. *Physiology and Molecular Plant Pathol*ogy, 61:267-271.

Akhtar, M. and Malik, A. (2000). Roles of organic soil amendments and soil organisms in the biological control of plant-parasitic nematodes: a review. *Bioresource Technology,* 74: 35–47.

Akiyama, K. (2007). Chemical identification and functional analysis of apocarotenoids involved in the development of arbuscular mycorrhizal symbiosis. *Bioscience, Biotechnology, and Biochemistry*, 71:1405–1414.

Akiyama, K., Matsuzaki, K. and Hayashi, H. (2005). Plant sesquiterpenes induce hyphal branching in arbuscular mycorrhizal fungi. *Nature*, 435: 824–827.

Anonymous, (1999). *The Ayurvedic Pharmacopoeia of India, Govt. of India*, First edition, Part I, Vol II, 158- 159.

Ashok, P. K., Upadhyaya, K. (2012). Tannins are astringent. *Journal of Pharmacognosy and Phytochemistry*, 1:45-50.

Bagayoko, M., George, E., Romheld, V. and Buerkert, A. (2000). Effects of mycorrhizae and phosphorus on growth and nutrient uptake of millet, cowpea and sorghum on a West African soil. *Journal of Agriculture Science,* 135: 399–407.

Bais, H. P., Park, S. W., Weir, T. L., Callaway, R. M. and Vivanco, J. M. (2004). How plants communicate using the underground information superhighway. *Trends Plant Science,* 9: 1360–1385.

Bais, H. P., Weir, T. L., Perry, L. G., Gilroy, S. and Jorge, M. (2006). The role of root exudates in rhizosphere interactions with plants and other organisms. *Annual Review of Plant Biology,* 57: 233–266.

Baleroni, C. R. S., Ferrarese, M. L. L., Souza, N. E. and Ferrarese-Filho, O. (2000). Lipid accumulation during canola seed germination in response to cinnamic acid derivatives. *Biologia Plantarum,* 43: 313–316.

Bartosz, G. (2004). *Druga twarz tlenu* [*The second face of oxygen*]. PWN, Warszawa, Poland.

Bavaresco, L., Mattivi, F., De Rosso, M. and Flamini, R. (2012). Effects of elicitors, viticultural factors, and enological practices on resveratrol and stilbenes in grapevine and wine. *Mini-Reviews in Medicinal Chemistry*, 12: 1366–1381.

Bertin, C., Yang, X. H. and Weston, L. A. (2003). The role of root exudates and allelochemicals in the rhizosphere. *Plant Soil*, 256: 67–83.

Bolwell, G. P. (1999). Role of active oxygen species and NO in plant defence responses. *Current Opinin Plant Biology,* 2:287-294.

Burghardt, F., Fiedler, K. and Proksch, P. (1997). Uptake of flavonoids from Vicia villosa (fabaceae) by the lycaenid butterfly, polyommatus icarus (lepidoptera: lycaenidae). *Biochemical Systematic and Ecology,* 25: 527-536.

Burghardt, F., Knüttel, H., Becker, M. and Fiedler, K. (2000). Flavonoid wing pigments increase attractiveness of female common blue (Polyommatus icarus) butterflies to mate-searching males. *Naturwissenschaften*, 87: 304-307.

Burghardt, F., Proksch, P. and Fielder, K. (2001). Flavonoid sequestration by the common blue butterfly Polyommatus icarus: quantitative intraspecific variation in relation to larval hostplant, sex and body size. *Biochemical Systematic and Ecology,* 29: 875-889.

Butler, L. G. (1982). Relative degree of polymerization of sorghum tannin during seed development and maturation. *Journal of Agricultural and Food Chemistry*, 30: 1090-1094.

Chen, C.-L. and Chang, H.-M. (1985). Chemistry of lignin biodegradation, in: *Biosynthesis and Biodegradation of Wood Components*, T. Higuchi, ed., Academic Press, Orlando, FL, pp. 535-556.

Coelho, J. V. and Majolo, F. M. (1993). Evolution of phenolic compounds and condensed tannins (proanthocyanidins) during the development of bean seeds (*Phaseolus vulgaris* L.). *Archives of Latinoamerican Nutrition*, 43: 61–65.

Constabel, C. P., Yip, L., Patton, J. J. and Christopher, M. E. (2000). Polyphenol oxidase from hybrid poplar. Cloning and expression in response to wounding and herbivory. *Plant Physiology,* 124: 285-295.

Corcuera, L. J. (1993). Biochemical basis for the resistance of barley to aphids. *Phytochemistry*, 33 (4):741-747.

Cushnie, T. T. P. and Lamb, A. J. (2005). Antimicrobial activity of flavonoids. *The International Journal of Antimicrobial Agents*, 26: 342–356.

Dakora, F. D. (2003). Defining new roles for plant and rhizobial molecules in sole and mixed plant cultures involving symbiotic legumes. *New Phytologist,* 158: 39–49.

Dakora, F. D. and Phillips, D. A. (2002). Root exudates as mediators of mineral acquisition in low-nutrient environments. *Plant Soil*, 245: 35–47.

Dao, L. and Friedman, M., (1994). Chlorophyll, chlorogenic acid, glycoalkaloid, and protease inhibitor content of fresh and green potatoes. *Journal of Agricultural and Food Chemistry,* 42: 633-639.

Deborah, S. D., Palaniswami, A., Vidhyasekaran, P. and Velazhahan, R. (2001). Time-course study of the induction of defense enzymes, phenolics and lignin in rice in response to infection by pathogen and non-pathogen. *Journal of Plant Disease and Protection,* 108: 204-216.

Delserone, L. M., McCluskey, K., Matthews, D. E. and VanEtten, H. D. (1999). Pisatin demethylation by fungal pathogens and nonpathogens of pea: association with pisatin tolerance and virulence, *Physiology and Molecular Plant Patholo*gy, 55: 317-326.

Demidchik, V. (2015). Mechanisms of oxidative stress in plants: From classical chemistry to cell biology. *Environmental and Experimental Botany,* 109:212-228.

Denton, F. R. (1998). Beetle juice, *Science* 281: 1285.

Dermauw, W., Wybouw, N. and Rombauts, S. (2013). A link between host plant adaptation and pesticide resistance in the polyphagous spider mite Tetranychus urticae. *Proceedings of the National Academy of Sciences of the USA*, 110: E113–E122.

Dixon, R. A. (2001). Natural products and plant disease resistance. *Nature,* 411: 843-847.

Ehrlich, P. R. and Raven, P. H. (1964). Butterflies and Plants, a Study in Coevolution. *Evolution*, 18: 586-608.

Felton, G. W., Donata, K., Del Vecchio, R. J. and Duffey, S. S. (1989). Activation of plant foliar oxidases by insect feeding reduces nutritive quality of foliage for noctuid herbivores. *Journal of Chemical Ecology,* 15: 2667-2694.

Ferrazzano, G. F., Amato, I., Ingenito, A., Natale De, A. and Pollio, A. (2009). Anti-cariogenic effects of polyphenols from plant stimulant beverages (cocoa, coffee, tea). *Fitoterapia*, 80: 255–262.

Friend, J. (1985). *The Biochemistry of Plant Phenolics*, Van Sumere, C. F. and Lea, P. J. (Eds), Clarendon Press, Oxford, 367.

Gill, S. S. and Tuteja, N. (2010). Reactive oxygen species and antioxidant machinery in abiotic stress tolerance in crop plants. *Plant Physiology and Biochemistry,* 48:909-930.

Gray, E. J. and Smith, D. L. (2005). Intracellular and extracellular PGPR: commonalities and distinctions in the plant–bacterium signaling process. *Soil Biology and Biochemistry,* 37: 395–412.

Halvorson, J. J., Gonzalez, J. M., Hagerman, A. E. and Smith, J. L. (2009). Sorption of tannin and related phenolic compounds and effects on soluble-N in soil. *Soil Biology and Biochemistry,* 41: 2002–2010.

Hammerschmidt, R. (1999). Phytoalexins: What have we learned after 60 years? *Annual Review of Phytopathol*ogy, 37: 285-306.

Hammerschmidt, R. and Nicholson, R. L. (2001). A survey of plant defense responses to pathogens, in: *Inducible plant defenses against pathogens and herbivores: Biochemistry, Ecology, and Agriculture. The American Phytopathological Society,* St. Paul, MN, pp 57-71.

Harborne, J. B. (1980). Secondary Plant Products, Bell, E. A. and Charlwood, B. W. (Eds), *Enciclopedia of Plant Physiology*, New Series, Vol. 8. Springer-Verlag, Berlin, 329.

Harborne, J. B. and Williams, C. A. (2000). Advances in flavonoid research since 1992. *Phytochemistry,* 55: 481–504.

Hartley, R. D. and Harris, P. J. (1981). Phenolic constituents of the cell walls of dicotyledons. *Biochemical Systematic and Ecology,* 9: 189–203.

Haruta, M., Pedersen, J. A. and Constabel, C. P. (2001). Polyphenol oxidase and herbivore defense in trembling aspen (*Populus tremuloides*): cDNA cloning, expression and potential substrates. *Physiologia Plantarum,* 112: 552-558.

Hättenschwiler, S. and Vitousek, P. M. (2000). The role of polyphenols in terrestrial ecosystem nutrient cycling. *Trends Ecology and Evolation,* 15: 238–242.

Hedin, P. A., Jenkins, J. N., Thompson, A. C., McCarty, J. C., Smith, D. H., Parrot, W. L. and Shepherd, R. L. (1988). Effects of Bioregulators on Flavonoids, Insect Resistance, and Yield of Seed Cotton. *Journal of Agriculture and Food Chemistry,* 36: 1055-1061.

Ikonen, A., Tahvanainen, J. and Roininen, H. (2001). Chlorogenic acid as an antiherbivore defence of willows against leaf beetles. *Entomologia Experimentalis et Applicata,* 99: 47-54.

Karou, D., Dicko, M. H., Simpore, J. and Traore, A. S. (2005). Antioxidant and antibacterial activities of polyphenols from ethnomedicinal plants of Burkina Faso. *African Journal of Biotechnology,* 4: 823–828.

Kefeli, V. I., Kalevitch, M. V. and Borsari, B. (2003). Phenolic cycle in plants and environment. *Journal of Cell Molecular Biology,* 2: 13–18.

Kojima, M. and Kondo, T. (1985). An enzyme in sweet potato root which catalyzes the conversion of chlorogenic acid, 3-caffeoylquinic acid, to isochlorogenic acid, 3,5-dicaffeoylquinic acid. *Agricultural and Biological Chemistry,* 49:2467-2469.

Koornneef, A. and Pieterse, C. M. J. (2008). Cross talk in defense signaling. *Plant Physiology,* 146: 839–844.

Kozłowska, M. and Konieczny, G. (2003). *Biologia odporności roślin na patogeny i szkodniki* [*Biology of plant resistance to pathogens and pests*]. Wyd. Akademii Rolniczej im. A. Cieszkowskiego, Poznan, Poland.

Kraus, T. E. C., Dahlgren, R. A. and Zasoski R. J. (2003). Tannins in nutrient dynamics of forest ecosystems – a review. *Plant Soil*, 256: 41–66.

Langcake, P., Irvine, J. A. and Jeger, M. J. (1981). Alternative chemical agents for controlling plant disease. *Philosophical transactions of the Royal Society of London. Series B, Biological sciences,* 295: 83–101.

Lattanzio, V., Lattanzio, V. M. T. and Cardinali, A. (2006). Role of phenolics in the resistance mechanisms of plants against fungal pathogens. *Phytochemistry,* 6: 23-67.

Lattanzio, V., Terzano, R., Cicco, N., Cardinali, A., Di Venere, D. and Linsalata, V. (2005). Seed coat tannins and bruchid resistance in stored cowpea seeds. *Journal of the Science of Food and Agriculture,* 85: 839–845.

Lo, S. C. C. and Nicholson, R. L. (1998). Reduction of light-induced anthocyanin accumulation in inoculated sorghum mesocotyls: Implications for a compensatory role in the defense response. *Plant Physiology*, 116: 979-989.

Lo, S. C. C., Hipskind, J. D. and Nicholson, R. L. (1999*).* cDNA cloning of a sorghum pathogenesis-related rotein (PR-10) and differential expression of defense-related genes following inoculation with *Cochliobolus heterostrophus* or *Colletotrichum sublineolum. Molecular Plant-Microbe Interactions*, 12*:* 479–489.

Lodhi, M. A. K., Bilal, R. and Malik, K. A. (1987). Allelopathy in agroecosystems: wheat phytotoxicity and its possible roles in crop rotation. *Journal of Chemical Ecology,* 13: 1881–1891.

Lu, H. (2009). Dissection of salicylic acid-mediated defense signaling networks. *Plant Signaling and Behaviour,* 4: 713–717.

Madhan, S. S. R., Girish, R., Karthik, N., Rajendran, R. and Mahendran, V. S. (2009). Allelopathic effects of phenolics and terpenoids extracted from *Gmelina arborea* on germination of Black gram (*Vigna mungo*) and Green gram (*Vigna radiata*). *Allelopathy Journal,* 23: 323–332.

Martens S., Preuss A. and Matern U. (2010). Multifunctional flavonoid dioxygenases: flavonol and anthocyanin biosynthesis in *Arabidopsis thaliana* L. *Phytochemistry,* 71: 1040-1049.

Matheis, G. and Whitaker, J. R. (1984). Modification of proteins by polyphenol oxidase and peroxidase and their products. *Journal of Food Biochemistry,* 8: 137-162.

Mert-Türk, F. (2002). Phytoalexins: defense or just a response to stress? *Journal of Cell Molecular Biology,* 1: 1–6.

Moerschbacher, B. M., Noll, U. M., Flott, B. E. and Reisener, H. J. (1988). Lignin biosynthetic enzymes in stem rust infected resistance and susceptible near-isogenic wheat lines. *Physiology and Molecular Plant Pathology*, 33: 33- 46.

Morris, S. W., Vernooij, B., Titatarn, S., Starrett, M., Thomas, S., Wiltse, C. C., Frederiksen, R. A., Bhandhufalck, A., Hulbert, S. and Uknes, S. (1998). Induced resistance responses in maize. *Molecular Plant–Microbe Interactions,* 11: 643–658.

Moulin, L., Munive, A., Dreyfus, B. and Boivin-Masson, C. (2001). Nodulation of legumes by members of the b-subclass of proteobacteria. *Nature*, 411:948–950.

Ndakidemi, P. A. and Dakora, F. D. (2003). Review: legume seed flavonoids and nitrogenous metabolites as signals and protectants in early seedling development. *Functional Plant Biology,* 30: 729–745.

Nicholson, R. and Hammerschmidt, R. (1992). Phenolic compounds and their role in disease resistance. *Annual Review of Phytopathology*, 30: 369

Nicholson, R. L. and Rahe, J. E. (2004). Apple scab and its management, in: *Fruit and vegetable diseases* Mukerji., K. G., ed., Kluwer Academic Publishers, Dordrecht, pp. 41-58.

Nicholson, R. L. and Wood, K. V. (2001). Phytoalexins and secondary products, where are they and how can we measure them? *Physiology Molecular Plant Pathol*ogy, 59: 63-69.

Nielsen, K. A., Gottfredsen, C. H., Buch-Pedersen, M. J., Ammitzbøll, H., Mattsson, O., Duus, J. Ø. and Nicholson, R. L. (2004). Anti-microbial flavonoid 3-deoxyanthocyanidins in *Sorghum bicolor* self-organize into spherical structures. *Physiology Molecular Plant Pathol*ogy, 65: 187-196.

Nishida, R., Ohsugi, T., Kokubo, S. and Fukami, H. (1987). Viposition stimulants of a Citrus-feeding swallowtail butterfly, *Papilio xuthus* L. *Experientia*, 43: 342–344.

Nowicka, B. and Kruk, J. (2013). Reaktywne formy tlenu w roślinach – więcej niż trucizna [Reactive oxygen species in plants - more than poison]. *Kosmos*, 62:583-596.

Perret, X., Staehelin, C. and Broughton, W. J. (2000). Molecular basis of symbiotic promiscuity. *Microbiology and Molecular Biology Reviews,* 64: 180–201.

Piasecka, A., Jedrzejczak-Rey, N. and Bednarek, P. (2015). Secondary metabolites in plant innate immunity: Conserved function of divergent chemicals. *New Phytologist,* 206: 948–964.

Puzanowska-Tarasiewicz, H., Starczewska, B. and Kuźmicka, L. (2008). Reaktywne formy tlenu. *Food and Chemical Toxicology*, 41:1007-1015.

Preisig, C. L., Matthews, D. E. and VanEtten, H. D. (1989). Purification and characterization of S-adenosyl-L-methionine: 6a-hydroxymaackiain 3-*O*methyltransferase from *Pisum sativum. Plant Physiology*, 91: 559-566.

Raa, J. and Overeem, J. C. (1968). Transformation reactions of phloridzin in the presence of apple leaf enzymes. *Phytochemistry,* 7: 721-731.

Raa, J. (1968) Polyphenols and natural resistance of apple leaves against *Venturia inaequalis, European Journal of Plant Pathology,* 74: 37-45.

Razal R. A., Ellis S., Singh S., Lewis N. G. and Tower G. H. N. (1996). Nitrogen recycling in phenylpropanoid metabolism. *Phytochemistry,* 41: 31–36.

Ridsdill-Smith, T. J., Jiang, Y. and Ghisalberti, E. L. (1995). A method to test compounds for feeding deterrence towards redlegged earth mite (Acarina: Perthaleidae). *Annals of Applied Biology*, 127: 593-600.

Scalbert, A. (1991). Antimicrobial properties of tannins. *Phytochemistry,* 30: 3875-3883.

Schardl, C. L., Leuchtmann, A. and Spiering, M. J. (2004). Symbiosis of grasses with seedborne fungal endophytes. *Annual Review Plant Biology,* 55: 315–340.

Schönbeck, F. and Schroeder, C. (1972). Role of antimicrobial substances (tuliposiodes) in tulips attacked by *Botrytis spp. Physiology and Plant Pathology,* 2: 91-99.

Seneviratne, G. and Jayasinghearachchi, H. S. (2003). Mycelial colonization by bradyrhizobia and azorhizobia. *Journal of Bioscience,* 28: 243–247.

Simmonds, M. S. J. (2001). Importance of flavonoids in insect--plant interactions: feeding and oviposition. *Phytochemistry*, 56: 245-252.

Simmonds, M. S. J. (2003). *Flavonoid–insect interactions: recent advances in our knowledge. Phytochemistry,* 64 (1): 21-30.

Simmonds, M. S. J. and Stevenson, P. C. (2001). Effects of isoflavonoids from Cicer on larvae of Helicoverpa. *Journal of Chemical Ecology,* 27:965–977.

Snook, M. E., Widstrom, N. W., Wiseman, B. R., Gueldner, R. C., Wilson, R. L., Himmelsbach, D. S., Harwood, J. S. and Costello, C. E. (1994). New flavone C-glycosides from corn (*Zea mays* L.) for the control of the corn earworm (*Helicoverpa zea*), in: *Bio-Regulators For Crop Protection And Pest Control,* P. A. Hedin, ed., Symposium Series 557 of the American Chemical Society, Washington, DC, pp. 122-135.

Snyder, B. A. and Nicholson, R. L. (1990). Synthesis of phytoalexins in sorghum as a site-specific response to fungal ingress. *Science.* 248: 1637- 1639.

Southerton, S. G. and Deverall, B. J. (1990). Histochemical and chemical evidence for lignin accumulation during expression of resistance to leaf rust fungi in wheat. *Physiology and Molecular Plant Patholo*gy, 36: 483-494.

Starr, J. L., Newton, R. J. and Miller, F. R. (1984). Presence of dhurrin in sorghum root tissue and the effect of pathogenesis on hydrogen cyanide potential. *Crop Science*, 24: 739-742.

Steinkellner, S., Lendzemo, V., Langer, I., Schweiger, P., Khaosaad, T., Toussaint, J. P. and Vierheilig, H. (2007). Flavonoids and strigolactones in root exudates as signals in symbiotic and pathogenic plant–fungus interactions. *Molecules*, 12: 1290–1306.

Sticher, L., Mauch-Mani, B. and Metraux, J. P. (1997). systemic acquired resistance. *Annual Review of Phytopathoogy*, 35: 235-270.

Taylor, L. P. and Grotewold, E. (2005). Flavonoids as developmental regulators. *Current Opinin of Plant Biology,* 8: 317–323.

Tosh, C. R., Powell, G., Holmes, N. D. and Hardie, J. (2003). Decision making by generalist and specialist aphids with the same genotype. *Journal of Insect Physiology*, 49: 659-669.

Treutter, D. (2005). Significance of flavonoids in plant resistance and enhancement of their biosynthesis. *Plant Biology,* 7: 581-591.

Treutter, D. (2010). Managing phenol contents in crop plants by phytochemical farming and breeding-visions and constraints. *International Journal of Molecular Sciences,* 11: 807–857.

Uknes, S., Winter, A., Delaney, T., Vernooij, B., Morse, A., Friedrich, L., Nye, G., Potter, S., Ward, E. and Ryals, J. (1993). Biological induction of systemic acquired resistance in Arabidopsis. *Molecular of Plant-Microbe Interactions,* 6: 692-698.

Urbanska, A., Tjallingii, W. F., Dixon, A. F. G. and Leszczynski, B. (1998). Phenol oxidising enzymes in the grain aphid's saliva. *Entomologia Experimentalis et Applicata,* 86: 197–203.

Van Loon, L. C. and Van Strien, E. A. (1999). The families of pathogenesisrelated proteins, their activities, and comparative analysis of PR-1 type proteins. *Physiology and Molecular Plant Patholo*gy, 55: 85–97.

Vit, P., Soler, C. and Tomás-Barberán, F. A. (1997). Profiles of phenolic compounds of *Apis mellifera* and *Melipona* spp. honeys from Venezuela. *Zeitschrift für Lebensmitteluntersuchung und -Forschung A,* 204: 43–47.

Ward, E. R., Uknes, S. J., Williams, S. C., Dincher, S. S., Wiederhold, D. L., Alexander, D. C., Ahl-Goy, P., Metraux, J. P. and Ryals, J. A. (1991). Coordinate gene activity in response to agents that induce systemic acquired resistance, *Plant Cell,* 3: 1085-1094.

Weir, T. L., Park, S. W. and Vivanco, J. M. (2004). Biochemical and physiological mechanisms mediated by allelochemicals. *Current Opinin of Plant Biology,* 7: 472–479.

Whipps, J. M. (2001). Microbial interactions and biocontrol in the rhizosphere. *Journal of Experimental Botany,* 52: 487–511.

Winkel-Shirley, B. (1998). Flavonoids in seeds and grains: physiological function, agronomic importance and the genetics of biosynthesis. *Seed Science and Research*, 8: 415-422.

Xuan, T. D., Shinkichi, T., Khanh, T. D. and Chung, I. M. (2005). Biological control of weeds and plant pathogens in paddy rice by exploiting plant allelopathy: an overview. *Crop Protection,* 24: 197–206.

Zagrobelny, M., Bak, S., Vinther Rasmussen, A., Jørgensen, B., Naumann, C. M. and Lindberg Møller, B. (2004). Cyanogenic glucosides and plant–insect interactions. *Phytochemistry,* 65: 293-306.

In: Agricultural Research Updates. Volume 27 ISBN: 978-1-53615-916-5
Editors: Prathamesh Gorawala et al.

Chapter 2

ASSESSMENT OF SUSTAINABILITY OF AGRO-ECOSYSTEMS IN BULGARIA

Hrabrin Bachev*, Bodjidar Ivanov and Dessislava Toteva
Institute of Agricultural Economics,
Sofia, Bulgaria

ABSTRACT

Ecosystem approach has been increasingly incorporated in the management and evaluation of sustainability levels. Despite the huge progress in the theory and practice of this new area, there is still no consensus on how to assess the sustainability of agro-ecosystems due to diverse understandings, approaches, methods, employed data, etc. In Bulgaria there are practically no deep studies on the sustainability level of diverse agro-eco-systems.

This chapter tries to fill the blank and assesses the sustainability level of agro-ecosystems of different type in Bulgaria. Firstly, a holistic hierarchical framework for assessing integral, economic, social and

---

* Corresponding Author's Email: hbachev@yahoo.com. This study was supported by the Bulgarian Science Fund (Project NoДНТС/Китай, 16.12.2016).

ecological sustainability of agro-ecosystems in Bulgaria is suggested, including 17 principles, 35 criteria, and 46 indicators and reference values. After that, an assessment is made on overall sustainability and its aspects of large (agro)ecosystems in North-Central, South-Eastern, South-Central and South-Western geographic regions, and particular of main and specific types of agro-ecosystems in the country - mountainous, plain-mountainous, plain, riparian (Struma, Maritza, Yantra), southern Black Sea, mountainous area with natural constraints, non-mountainous area with natural constraints, protected areas and reserves, Western Thracian Plain, Middle Danube Plain, Dupnitsa and Sandansko-Petrich Valley, Sredna Gora Mountains and Western Rila Mountains. The assessment is based on first-hand information collected though in-depth interviews with the managers of "typical" farms in the respective agro-ecosystems.

The study has found out that there is a considerable differentiation in the level of integral sustainability in agricultural ecosystems of different types. Furthermore, there are substantial variations in the levels of economic, social and ecological sustainability of agro-ecosystems of different type, and the critical indicators enhancing or deterring the overall and the particular sustainability of individual agro-ecosystems. The results of the integral agricultural sustainability level of this study, based on the micro agro-ecosystem (farm) data, are similar to the previous assessment based on the aggregate sectoral (statistical, etc.) data. There are large differences in the impact of socio-economic, institutional, behavioral, international, natural, etc. factors and individual public policy instruments on the sustainability of farming enterprises of different types and agro-ecosystems.

Having in mind the importance of holistic assessments of this kind for improving agricultural sustainability, farm management and agricultural policies, they should be expended and their precision and representation increased.

Keywords: agro-ecosystem, sustainability, assessment, economic, social, ecological, Bulgaria

INTRODUCTION

The issue of assessment of the sustainability of agricultural systems of various type is among the most topical for last decades (Bachev, 2009, 2010, 2016, 2017, 2018; Bachev et al., 2016, 2017; Candido et al., 2018;

FAO, 2013; Fuentes 2004; Hayati et al., 2010; Ikerd, 2015; Ivanov et al., 2009; Gliessman, 2016; Gemesi, 2007; Gitau et al., 2009; Jalilian, 2012; Irvin et al., 2016; Lopez-Ridauira et al. 2002; Rezear et al, 2018; Sauvenier et al., 2005; Terziev et al., 2018; Todorova and Treziyska, 2018; VanLoon et al. 2005; Zvyatkova and Sarov, 2018).

Agro-ecosystems are ecosystems associated with agricultural (farming) activity and according to their specific characteristics and levels of analysis the borders of an individual agro-ecosystem could be a part of a separate farm (e.g., a cultivated parcel, a meadow, a pond), located in numerous farms, or most commonly cover a larger region(s) of a country or beyond. Moreover, the individual agro-ecosystem could include, be a part, or overlap with other ecosystems - dryland, mountain, coastal, urban, etc.

In recent years an "ecosystem approach" has been increasingly incorporated in the management and evaluation of sustainability levels (Bachcv and Treziev, 2017, 2018; Belcher, 1999; Bohlen and House, 2009; Hanna et al., 2016; MEA, 2005; De Oliveira, 2018; Ramírez-Carrillo et al., 2018; Oelbermann, 2014; Sidle et al., 2013). Despite the enormous progress in the theory and practice of this new evolving area, there is still no consensus on how to assess the sustainability of agro-ecosystems due to diverse understandings, approaches, methods, employed data, etc.

In Bulgaria comprehensive sustainability assessments are mostly on sectoral (Bachev et al., 2017) or farm (Bachev, 2017; Bachev and Treziev, 2017) levels, while there is practically no in-depth study on sustainability of agro-ecosystems.

The goal of this paper is to assess the sustainability level of agro-ecosystems of different types in Bulgaria.

Methodology and Data

In order to assess the sustainability level of agro-ecosystems in Bulgari, a hierarchical system is developed, including 17 principles, 35 criteria, and 46 indicators and reference values. Principles are the highest hierarchical level, associated with the "universal" functions of agricultural system and represent the state of sustainability in 3 main pillars (aspects) of sustainability - economic, social, and ecological. Criteria represent a resultant state of realization of the relevant principle. Indicators are quantitative and qualitative variables of different types (behavior, activity, input, effect, impact), which can be assessed, allowing the measurement of compliance with particular criteria. Reference values are the desirable levels (absolute, relative, qualitative) for each indicator according to the specific conditions of each agro-ecosystem, which assist the assessment giving guidance for achieving (maintaining, improving) the sustainability.

We have examined the available academic research, official documents, and experience in Bulgaria and other countries, and have carried out numerous consultations with leading national and international experts in the area. On this basis, a system that includes principles, criteria, indicators, and reference values, relevant to contemporary conditions in Bulgaria, has been formulated. An expert panel was set up with ten leading experts in the country, which have discussed and evaluated the importance of the proposed principles, criteria, indicators, and reference values, and selected the most appropriate to the contemporary conditions in Bulgaria (Table 1). A number of criteria were used in selecting indicators: relevance to reflecting aspects of sustainability; discriminatory power in time and space; analytical soundness; intelligibility and synonymy; measurability, governance and policy relevance; and practical applicability (Sauvenier et al., 2005).

In Bulgaria, such as in the most countries, there are no official data for calculating the socio-economic and (some parts of) ecological indicators at agro-ecosystem level. Agro-ecosystems are the ecosystems associated with the farming activity and the individual farm is the first level for governing the ~~of~~ agrarian sustainability (Bachev, 2018).

Table 1. System of principles, criteria, indicators, and reference values for assessing sustainability level of agro-ecosystems in Bulgaria

Principles	Criteria	Indicators	Reference values
Economics aspect			
Financial stability	Reducing dependence on subsidies	Share of direct payments in Gross Value Added	Experts estimate/Trend
	Sufficient liquidity	Ratio of overall liquidity	Experts estimate/Trend
		Ratio of quick liquidity	Experts estimate/Trend
	Minimizing dependence on external capital	Share of owned in total capital	Experts estimate/Average for the sector
Economic effectiveness	Positive or high profitability	Cost - effectiveness	Experts estimate/Average for the sector
		Profitability of capital	Experts estimate/Average for the sector
	Maximize or increase labour productivity	Labour productivity	Experts estimate/Average for the sector
	Maximize or increase land productivity	Productivity of land	Experts estimate/Average for the sector
	Maximize or increase livestock productivity	Livestock productivity	Experts estimate/Average for the sector
Competitiveness	Support or increase of marketed output	Share of marketed output	Experts estimate/Trend
	Support or increase of sales	Sales growth in the last 3 years	Experts estimate/Trend
Adaptability to economic environment	Sufficient adaptability to market environment	Ratio of gross income to fixed costs	Experts estimate/Trend
	High investment activity	Investment growth	Average for the sector/Trend

Table 1. (Continued)

Principles	Criteria	Indicators	Reference values
Social aspect			
Welfare of employed in agriculture	Equality of income with other sectors	Ratio of farm income to the average income in the region	Experts estimate/Trend
	Fair distribution of income in agriculture	Ratio of payment of hired labour in the farm to average income in the region	Average for the sector/Trend
	Sufficient satisfaction from farm activity	Degree of satisfaction from farm activity	Farmers assessment
	Satisfactory working conditions	Correspondence to official norms	Official norms
Conservation of farming	Preservation of the number of family farms	Existence of a heritor ready to take over of the farm	Experts estimate/Trend
		Number of family workers	Experts estimate/Trend
		Age of the manager	Farmers assessment/Trend
	Increasing the knowledge and skills	Level of participation in the training programs	Experts estimate/Trend
		Level of education of the manager	Experts estimate/Trend
	Maintaining and increasing of agrarian education	Number of employed with special agricultural education	Experts estimate/Trend
Gender equality	Equality in men-women relations	Degree of participation of women in farm management	Half/Trend

Principles	Criteria	Indicators	Reference values
Social capital	Participation in professional associations and initiatives	Number of participations in professional associations and initiatives	Experts estimate
		Level of hired labour membership in labour unions	Experts estimate/Trend
	Participation in public management	Public position	Experts estimate/Trend
	Contribution to the development of regions and communities	Participation in local initiatives	Experts estimate/Trend
Adaptability to the social environment	Sufficient ability to respond to the ceasing farming activity and the demographic crisis	Vacant job positions in the farms to the total number of employed	Experts estimate/Trend
Ecological aspect			
Air quality	Maintaining and improving air quality	Growth of carbon emissions for the past three years	Trend
Land quality	Minimizing soil losses	Soil erosion index	Scientific norm/Trend
	Preservation and improvement of soil fertility	Amount of nitrogen fertilization	Scientific norm/Average for the sector
		Amount of potassium fertilization	Scientific norm/Average for the sector
		Amount of phosphorus fertilization	Scientific norm/Average for the sector
	Maintaining a balanced land use structure	Share of arable land (without fallow) in total agricultural areas	Scientific norm/Average for the sector
	Preservation of landscape features	Amount of area covering the requirements for "green" direct payments through maintaining landscape elements	Experts estimate/Trend

Table 1. (Continued)

Principles	Criteria	Indicators	Reference values
Water quality	Maintaining and improving water quality	Index of groundwater pollution	Scientific norm/Average for the sector
Effective energy consumption	Minimizing the use of conventional energy	Fuel consumption per unit area	Experts estimate/Average for the sector
		Cost of conventional electric energy per unit of gross output	Trend/Average for the sector
Biodiversity	Maintaining or enhancing natural habitats	Change in the number of habitats	Trend/Average for the sector
		Share of agricultural land in NATURA 2000 and other protected areas	Planed target Trend
	Preserving and improving the biodiversity	Number of cultivated plant species	Trend/Average for the sector
Animal welfare	Compliance with the principles of animal welfare	Level of compliance with the principles of animal welfare	Official norms
Implementation of organic production	Increasing the organic production	Share of areas under conversion or certified for organic production	Experts estimate/Trend
Adaptability to the environment	Sufficient adaptability to climate change	Variation in the yield of main crops	Average for the sector/ Trend
		Death rate in livestock farms	Average for the sector/ Trend

Source: Authors.

In order to assess the level of sustainability of agro-ecosystems in the country, in-depth interviews with the managers of 80 farms of different types and locations in 4 major regions of Bulgaria were held in 2017. The following criteria were used for the selection of areas for farm surveys (Map 1):

- major administrative and geographic regions - Eastern, Northern, Western and Southern Bulgaria respectively North-Central, South-Eastern, South-Central and South-Western administrative and geographic regions of the country representing distinctive large (agro)ecosystems;
- particular main types and specific (agro) ecosystems in the country - mountainous, plain-mountainous, plain, riparian (Struma, Maritza, Yantra), southern Black Sea, mountainous area with natural constraints, non-mountainous area with natural constraints, protected areas and reserves, Western Thracian Plain, Middle Danube Plain, Dupnitsa and Sandansko-Petrich Valley, Sredna Gora Mountains and Western Rila Mountains.

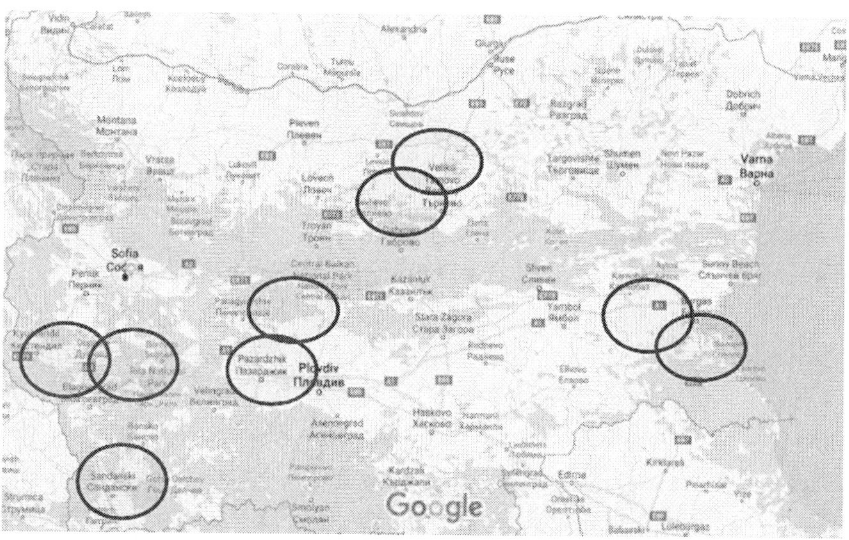

Source: Google maps.

Map 1. Map of Bulgaria and surveyed agro-ecosystems.

In order to identify the “typical” for the different regions of the country farms, the co-operation of the main associations of farmers (National Association of Grain Producers, National Union of Gardeners, Union of Breeders, etc.), state agencies (National Agricultural Advisory Service, Executive Agency for Vine and Wine, etc.), processing, bio-certification and service organizations, and local government is used. Farmers of different types were surveyed, covering the main types of farms in the regions concerned: different legal types of holdings - natural persons, sole traders, cooperatives, commercial companies, etc. .; farms of different sizes - mainly for self-sufficiency, with small size for the sector, with average size for the sector, with large sizes for the sector; farms of different production specialization - arable crops, vegetables, flowers and mushrooms, perennials, grazing livestock, pigs, poultry and rabbits, mixed crops and mixed livestock breeding; farms in specific geographic and ecological locations. From the farms originally identified for interviews only 5,61% were not interviewed, due to the extreme busyness, unwillingness to participate or other reasons.

During the surveys, the managers of the farms were aware with the objectives of the survey, they replied to the questions prepared in advance and discussed the main problems and challenges of sustainable agriculture in the farms and eco-systems. The duration of the interview with each participant was from several hours to a whole day, and in many cases additional meetings and phone calls were conducted to refine and supplement the answers.

The survey includes many questions in 5 major areas: general characteristic of farms; primary information for calculating economic indicators for agrarian sustainability at agro-eco-system level; primary information for calculating social indicators for agrarian sustainability at agro-eco-system level; primary information for calculating environmental indicators for agricultural sustainability at agro-eco-system level; impact of diverse socio-economic, policies, behavioral, personal, etc. factors on farmers actions for improving agricultural sustainability and its various aspects.

After that diverse quantitative and qualitative levels for each indicator are transformed into a unitless index of sustainability (ISi). After that, the integral index for a particular criterion (SI(c)), principle (SI(p)), and aspect of sustainability (SI(a)), and the integral sustainability index (SI(o)) for each surveyed farm is calculated, applying equal weight for each indicator in a particular criterion, of each criterion in a particular principle, and each principle in every aspect of sustainability.

The arithmetic averages of the indices of composite indicators, criteria and principles are calculated by the following formulas:

$SI(c) = \sum SI(i)/n$ n - number of indicators in a particular criterion;
$SI(p) = \sum SI(c)/n$ n - number of criteria in a particular principle;
$SI(a) = \sum SI(p)/n$ n - number of principles in a particular aspect,
$SI(o) = \sum SI(a)/3$

The composite sustainability index of a particular agri-ecosystem is an arithmetic average of the indices of relevant farms belonging to that agro-ecosystem.

For assessing the level of sustainability of agro-ecosystems the following scale, defined by the experts, is used:

Index range 0,85-1 for a high level of sustainability;
Index range 0.50-0,84 for a good level of sustainability;
Index range 0,25-0,49 for a satisfactory level of sustainability;
Index range 0,12-0,24 for an unsatisfactory level of sustainability;
Index range 0-0,11 for non-sustainable.

General Characteristic of the Questionnaire for the Farms

The survey was conducted in the period April-November 2017 and covered 80 farmers from five administrative districts of the country - Pazardjik, Plovdiv, Kjustendil, Blagoevgrad, Bourgas and Veliko Tarnovo (Table 2).

Table 2. Geographical and ecological location of surveyed agricultural holdings (number)

Location of farms	North-Central Region	South-western Region		South-Central Region		South-eastern region	General number * and %
	Veliko Tarnovo	Kjustendil	Blagoevgrad	Pazardzhik	Plovdiv	Bourgas	
Mostly plane area	2	4	4	14	0	8	80
Plane-mountain area	8	4	2	8	2	6	37,5
Mostly mountain area	0	6	2	4	6	0	22,5
Land in protected areas and territories	0	0	0	0	2	4	7,5
Mountain area with natural restrictions	2	6	0	4	0	2	17,5
Non-mountainous area with natural restrictions	0	0	2	2	0	0	5
Western Thracian Lowland	0	0	0	22	0	0	27,5
Middle Danube Plain	6	0	0	0	0	0	7,5
Dupnitsa valley	0	4	0	0	0	0	5
Sandanski-Petrich valley	0	0	6	0	0	0	7,5
The valley of the Maritsa river	0	0	0	14	0	0	17,5
The valley of the Yantra river	6	0	0	0	0	0	7,5
The valley of the Struma River	0	4	6	0	0	0	12,5
South-Black Sea	0	0	0	0	0	8	10
Sredna Gora mountain	0	0	0	6	6	0	15
Western Rila mountain	0	4	2	0	0	0	7,5
Total number	10	14	8	26	8	14	80*
Share of all holdings (%)	12,5	17,5	10	32,5	10	17.5	100

Source: Survey with managers of farms, 2017.

Table 3. Legal status, sizes and production specialization of the surveyed agricultural farms (number)

Type of farms	North-Central Region	South-western Region		South-Central Region		South-eastern Region	Share in Total number (%)
	Veliko Tarnovo	Kjustendil	Blagoevgrad	Pazardzhik	Plovdiv	Bourgas	
Legal person	6	6	2	6	6	4	37,5
Sole trader	2	4	4	6	0	0	20
Cooperative	2	2	0	4	0	4	15
Commercial company, etc.	0	2	2	10	2	6	27,5
Companies, mostly for self-sufficiency	0	2	0	0	4	0	7,5
Companies, rather small for the industry	4	6	2	14	2	2	37,5
Companies, average for the industry	4	4	4	10	0	6	35
Companies, big for the industry	4	0	2	2	2	6	20
Field crops	2	2	0	2	0	4	12,5
Vegetables, flowers and mushrooms	0	2	2	4	0	0	10
Perennial plants	4	0	4	6	2	4	25
Grazing animals	2	0	0	2	2	0	7,5
Pigs, poultry and rabbits	0	2	0	2	0	0	5
Mixed crop-livestock farms	2	4	2	4	4	4	25
Mixed crop farms	0	2	0	6	0	2	12,5
Mixed livestock farms	0	2	0	0	0	0	2,5

Source: Survey with managers of farms, 2017.

The majorities of the surveyed holdings are unregistered farms of individuals, mostly small in size, and specialized in mixed crop-livestock farming and perennial crops (Table 2). Most of the studied farms are located in South Central and South-West geographical and administrative regions, and in mostly plane and plane-mountain areas of the country. One quarter of the surveyed farms is in the Thracian Lowland. Each fifth is located in valleys of different kind - Danube plain, Dupnitsa valley and Sandanski-Petrich valley. In riverside ecosystems of different types (Maritsa, Struma and Yantra) there are about 36% of the farms surveyed and in the seaside area - every tenth farm.

The owners or managers of the majority of surveyed farms are men and in active working age, from 41 to 65 years. Such gender and age structure of managers (owners) will manage the majority of Bulgarian farms in the near 10-15 years and will contribute to one or other level of their sustainability. The majority of respondents are between 56 and 65, which is an indicator of both their life and professional experience and an alarming sign for the aging of the employed persons in our agriculture.

Most of the surveyed farms have a relatively long life - over 15 years and only 10% are with a short development period - from 2 to 5 years. This is an indicator that the majority of farms have sufficient effective management experience and sustainability. Most of the farmers surveyed indicate that the period they are taking care of improving the sustainability of the farm is over 6 years, the majority of them are in the group with long experience over 15 years. There is a correlation between the duration of the existence of the farms and the period during which the farms take care to improve their sustainability. Moreover, with the increase in the duration of the existence of the farm, the proportion of farms with an effective care to improve their sustainability increases. All this shows that the practical problem of "agricultural sustainability" is not new. However, the question is whether farms know and to what extent they respect the principles of sustainable agriculture.

The knowledge of the main socio-economic and environmental challenges and the basic principles of sustainable agriculture is the basis for effective management of agricultural sustainability. Our large-scale survey has found that according to the majority of farms in the country, they are located in areas with "normal" economic, social and environmental problems. However, a significant part of them is in the areas with "big" or "extreme" economic, social and environmental challenges. One third of the managers say that their farm is located in an area with "small" or "no" ecological problems, while the share of farms with similar economic and social problems is smaller. The share of managers who are not familiar with the character or cannot assess the level of socio-economic and environmental problems in the area where their farm is located is not low. The greatest concern is farmers' competence with regard to the ecological problems in the area, followed by social and economic challenges.

Our study found that the majority of the managers of the surveyed farms know "well" and "very well" the principles of economic, social and environmental sustainability. At the same time, a large proportion of farmers recognize that their knowledge of the principles of social and environmental sustainability is "satisfactory" or lacking at all. The low lack of competence concerns almost the half of the holdings in terms of social sustainability principles, almost every third farm in terms of environmental sustainability and about one fifth of farms for economic sustainability.

Only a small proportion of the surveyed farms has increased its capacity for sustainable management by hiring a consultant, and this is all about getting to know the principles of environmental and economic sustainability. A relatively high (internal) potential for managing the different aspects of sustainability have cooperative farms, where everyone knows "well" or "very well" the principles of economic and social sustainability, and a significant part of them know the principles of environmental sustainability (Figure 6). At the same time, 16.67% of these farms "use a consultant" to improve their environmental sustainability competence.

All of the sole traders know well or very well the principles of economic sustainability and three-quarters of them - the principles of environmental sustainability. About 12% of these types of farms hire a consultant in order to improve the economic sustainability. The majority of sole traders also know well or very well the principles of social sustainability. However, 37.5% of them report that their knowledge about the principles of social sustainability is not good. The majority of commercial companies know well or very well the principles of economic and environmental sustainability, but only slightly more than half of them have a similar level of competence with respect to the principles of social sustainability. Every tenth of this type of farms also use an external consultant to enhance its environmental sustainability competence. Two thirds of individuals are highly competent in terms of economic sustainability principles, and 40% of them are also competent in terms of environmental sustainability. At the same time, nearly three quarters (73.33%) of this type of farms are not well aware of the principles of social sustainability.

The competence of sustainability principles grows together with the farm size and, as a rule, the larger farms are better acquainted with economic, social and environmental sustainability. At the same time, 7.69% of the medium-sized farms hire a consultant to increase their knowledge of economic sustainability and 15.38% of environmental sustainability. At the same time, it is worrying that none of the farms that are predominantly self-sufficient knows well the principles of economic, social and environmental sustainability. This group of producers represents a significant part of all farms in the country and is an important factor in improving the socio-economic and environmental sustainability of agriculture. There is also a differentiation of competence with respect to the principles of sustainability and depending on the production specialization of farms. In all categories of farms, a high level of knowledge of the principles of economic sustainability is typical of all or a majority of them. Exceptions are only the farms with crop growing specialization, where each second farm is not well aware with the principles of economic sustainability. Half of pig, poultry and rabbit farms also have a consultant to improve their competence in terms of economic sustainability.

The knowledge of the principles of ecological sustainability is high in farms specializing in field crops, perennial crops, mixed crops, mixed crops and grazing livestock, while in farms with other specialization the share of those with low ecological competence is significant. Each fifth of field crops' farms improves its ecological sustainability capacity by hiring a consultant, similar to 11.11% of those in perennial crops. The knowledge of the principles of the social sustainability is good in most of the farms specializing in field plants, mixed crop growing and perennial crops. For farms in other production specialization, the share of highly competence in social sustainability is low, and for farms with vegetables, flowers and mushrooms, and those of mixed livestock farming, their share is zero.

The farms located in predominantly plain and plain-mountain areas and those in non-mountainous areas with natural constraints have a better knowledge of the principles of economic, social and environmental sustainability. On the other hand, farms located in predominantly mountainous areas, in mountainous areas with natural constraints and those with landscapes in protected areas and territories have a relatively low competence in the principles of sustainability. Some of the farms located in mountainous regions improve their economic and ecological sustainability by employing a consultant - respectively 6.67% and 13.33% of all farms in this group.

Finally, all the farms surveyed in the South-East region know well or very well the economic, social and ecological principles of agrarian sustainability. The competence for the economic sustainability is high in most of the farms in the other studied regions of the country. Most of the farms in the North-Central region are well informed about the environmental sustainability, while in the South-West region they are a minority. Also, the knowledge of the social sustainability principles is not good at the majority of farms in the South-Central and South-West regions of the country. Consultants in order to improve the knowledge of sustainable agriculture use 13.5% and 6.25% of farms in the South-West and South-Central region in terms of ecological aspects and 6.25% of farms in the South Central Region in terms of economic sustainability. Therefore, in the future, greater efforts should be made in order to improve the farmers' competence in low-culture groups with regard to the principles of agricultural sustainability through training, counselling, advices, exchange of positive experiences, etc.

The competence about the principles of agrarian sustainability is a necessary but not a sufficient condition for its effective management. Due to incomplete knowledge and various other economic, technological, agronomic, behavioural, etc. reasons, and at different times, farmers do not always strictly apply the principles of sustainable agriculture. Our study has found that, according to the majority of farm managers, they comply "strict" or "good" principles of economic, ecological and social sustainability (Bashev 2016). However, a significant part of the farms respects the principles of social, economic and environmental sustainability only "satisfactory." Moreover, some farms point that they do not "follow" such principles (which reach 6% of the total number of farms in terms of social sustainability), or "only follow if there are sanctions" (up to 8% for the ecological sustainability).

The principles of agricultural sustainability are applied to the greatest extent in the general management of farms in cooperatives and commercial companies. Around 8% of cooperatives apply the principles of environmental sustainability only if there are sanctions. A comparatively smaller proportion of sole traders and natural persons apply the principles of social sustainability to a high degree. Many natural persons follow the principles of sustainable agriculture only if there are sanctions - 9% for environmental sustainability, 5% for economic sustainability and 5% for social sustainability. These data show that sanctions by the state, local authorities, owners, members, etc. generate economic behaviour to improve environmental sustainability in certain groups of farms such as cooperatives and natural persons.

The application of sustainability principles grows with farm sizes and as a rule, larger farms are better of economic, social and environmental sustainability. The compliance with the diversity of sustainability principles is most common among farms specializing in field plants, grazing livestock and mixed crop growing farms. However, the quoted study also found that for all groups of holdings, the proportion of those who respect well or strictly the principles of agricultural sustainability exceeds the proportion of those who know well or very well these principles. Therefore, the question is to which extend some of the farms apply effectively some principles that they themselves do not know well.

Overall Level of Sustainability in Analyzed Agro-Ecosystems

The multi-indicator assessment of agricultural sustainability level in the four analyzed regions shows that the integral indicator of overall sustainability is 0,58, which expresses a good sustainability level of agriculture (Figure 1). The biggest value has the indicator of economic sustainability (0,64), the social sustainability shows lower value (0,57) and the ecological sustainability is close to the unsatisfying value level (0,53). Therefore, the improvement of the last two indicators is crucial for maintaining the good agricultural sustainability of the country.

The analysis of private indexes on basic principles, criteria and indicators of the sustainability gives opportunity to identify components contributing for the levels of different aspects of agricultural sustainability in the country. The assessment ascertained that the ecological sustainability is relatively low due to the fact that the indicators for the principles "land quality" (0,44), "biodiversity" (0,38) and "organic production" (0,11) are low (Figure 2). Thus, the improvement of these low levels of above-mentioned principles is a factor for maintenance and raising of ecological and integral sustainability in the sector. Also it becomes clear that despite the relatively high integral economic sustainability, the indicator of adaptability to economic environment is relatively low (0,54) and critical for maintaining the reached level. Analogically, for the social sustainability improvement would contribute mostly the increase of low levels of indicators for the principles "farming conservation" (0,52), "gender equality" (0,40) and "social capital" (0,17).

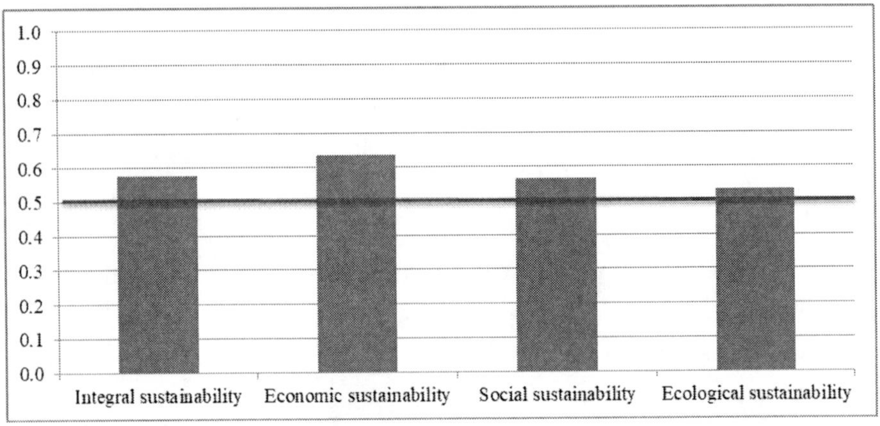

Source: Survey with managers of farms, 2017 and author's calculations.

Figure 1. Indicators of integral, economic, social and ecological sustainability of analyzed agri-ecosystems in Bulgaria.

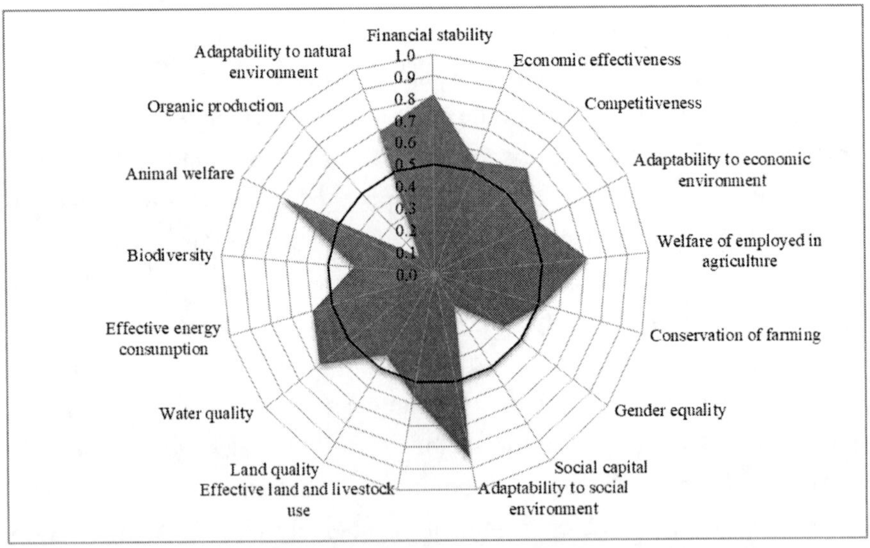

Source: Survey with managers of farms, 2017 and author's calculations.

Figure 2. Sustainability index according the main sustainability principles in analyzed agri-ecosystems in Bulgaria.

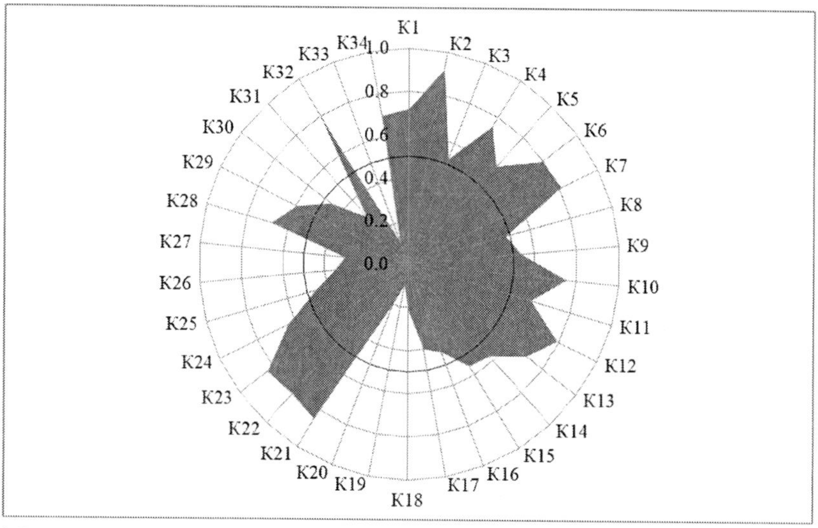

*K1-Decrease of dependence on subsidies; K2-Minimization of dependence on exterior capital; K3-Positive or high profitability; K4-Maximal or increasing labour productivity; K5-Maximal or increasing land productivity; K6-Maximal or increasing livestock productivity; K7-Conservation or increase of sold output share; K8-Conservation or increase of sales; K9-High investment activity; K10-Incomes parity with other sectors; K11-Equitable distribution of income in agriculture; K12-Sufficient satisfaction of farmer activity; K13-Satisfying labour conditions; K14-Keeping the number of family farms; K15-Knowledge and skills increase; K16-Conservation and improvement of agricultural education; K17-Equality of relations man-woman; K18-Participation in professional organizations and initiatives; K19-Participation in public management; K20-Contribution for the development of region and communities; K21-Sufficient potential for reaction to activity cession and to demographic crisis; K22-Keeping or increase of UAA size; K23-Keeping or increase of livestock number; K24-Minimization of soil losses; K25-Keeping and improvement of soil fertility; K26-Keeping of balanced land-use structure; K27-Protection of landscape elements; K28-Keeping and improvement of water quality; K29-Minimization of conventional energy use; K30-Keeping and improvement of natural biodiversity; K31-Keeping and improvement of cultural biodiversity; K32-Implementation of principles of animal welfare; K33-Organic production increase; K34-Sufficient adaptability to climatic changes.

Source: Survey with managers of farms, 2017 and author's calculations.

Figure 3. Sustainability index according the main criteria* in analyzed agri-ecosystems in Bulgaria.

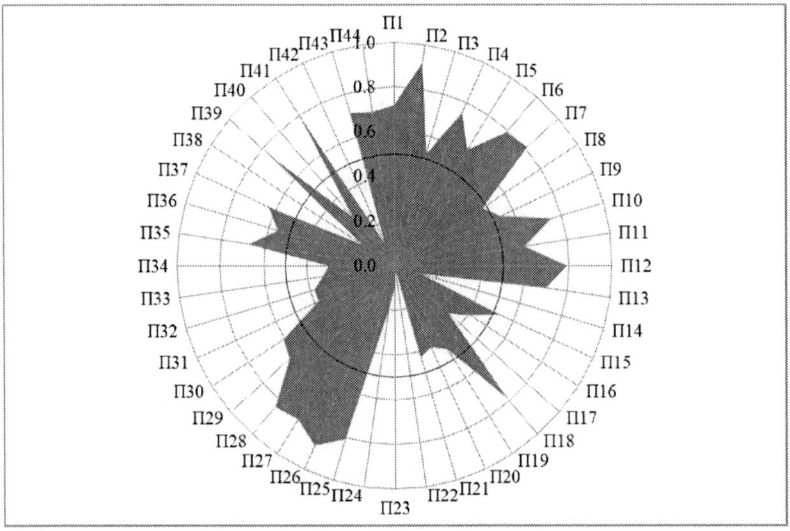

*П1-Direct payments in the net income; П2-Share of own capital in the total one; П3-Profit/production costs; П4-Labour productivity; П5-Land productivity; П6-Livestock productivity; П7-Share of sold production in the total one; П8-Sales growth in the last three years; П9-Investments growth in last 5 years; П10-Net farmer's income/average income in the region; П11-Payment of hired labour/average income in the region; П12-Degree of satisfaction from farmer's activity; П13-Degree of compliance to normative labour conditions; П14-Presence of a family member ready to take the farm; П15-Number of family members working in the farm; П16-Age of manager; П17-Participation of training programs in the last 3 years; П18-Education level of manager; П19-Share of occupied with special agricultural education/qualification; П20-Degree of participation of women in the farm management; П21-Number of participation in professional organizations and initiatives; П22-Share of hired workers, members of trade unions; П23-Public positions occupied from the farmer, manager and owner; П24-Participation in local initiatives; П25-Share of non-occupied permanent work positions in the total number of employed; П26-Share of non-occupied seasonal work positions in the total number of employed; П27-Change of UAA in last 5 years; П28-Change of livestock number in last 5 years; П29-Soil erosion; П30-Compliance of nitrate fertilization to norms; П31-Compliance of potassium fertilization to norms; П32-Compliance of phosphorus fertilization to norms; П33-Share of arable land in the total UAA; П34-Keeping the practices of landscape maintenance; П35-Degree of pollution of underground waters with nitrates; П36-Level of fuel consumption; П37-Level of electricity consumption; П38-Presence of protected species on the farm territory; П39-Natural biodiversity protection; П40-Number of cultural species; П41-Respecting of animal welfare norms; П42-Implementation of principles for organic production; П43-Yield variation of main crops for 5 years; П44-Percentage of mortality of livestock for 5 years.

Source: Survey with managers of farms, 2017 and author's calculations.

Figure 4. Indicators* for sustainability in analyzed agro-ecosystems in Bulgaria.

The profound analysis according different criteria and indicators gives opportunity for detailed analysis of elements contributing for/or decrease the agricultural sustainability level. For example, the low levels of ecological sustainability are determined by the low criteria "conservation and improving of soil fertility" (0,46); "balanced land use structure maintenance" (0,35; "landscape elements conservation" (0,30); "natural biodiversity maintenance and improvement" (0,46); "cultural biodiversity maintenance and improvement" (0,29) and "organic production increase" (0,11) (Figure 3). The unsatisfying levels according these criteria for ecological sustainability are (pre)determined of low levels of indicators for eco-sustainability, as: insufficient conformity of norms for fertilization with potassium (0,38) and phosphorus (0,38), high share of arable land in the total agricultural land (0,33), low degree of compliance with practices for landscape conservation (0,3), insufficient protected species on farms' territory (0,18), limited number of cultural species in farms (0,29) and low degree of application of organic production principles (0,11) (Figure 4).

The social sustainability in agriculture is usually decreased by: lack of family member, ready to continue the farm work (for individual and family farms) (0,13), elderly age of managers and farm owners (0,41), insufficient participation in training programs in the last years (0,33), low share of employed with special agricultural education and qualification (0,44), insufficient participation of women in the farm management (0,4), low participation of farms in professional organizations and initiatives (0,43), lack of membership of hired workers in trade unions (0), weak participation in the public governance from the side of farmers, managers and owners (0,1), and insufficient involvement of farms in local initiatives (0,2).

Critical for the keeping and improvement of the sector's economic sustainability are the increase of production profitability (0,52) and the keeping and increase of sales (0,48). The low levels of indicators for sustainability show also the specialized areas for agricultural sustainability improvement through adequate change of farms strategies and/or of public policies in relation to the sustainable development of the sector, of different sub-sectors, ecosystems and farms types. On the other hand, the high levels of some indicators express the absolute and relative advantages of Bulgarian agriculture regarding the sustainable development. On the actual stage they are expressed in: high share of own capital in the total capital of farms (0,92), high share of sold production in the total output (0,81), lower share of non-occupied permanent (0,81) and seasonal (0,88) work places in the total number of employed, increase of UAA (0,82) and livestock number (0,84) in the last years and respect of norms for animal welfare (for the livestock breeding farms) (0,8).

Level of Agricultural Sustainability in the Main Types of Agro-Ecosystems

Our assessment determined that there is a considerable differentiation of the level of integral and aspect sustainability in agricultural ecosystems main types (Figure 5). The highest integral sustainability has the agriculture in the plane regions (0,63), which have also the highest economic sustainability, with the ecosystems in protected zones and territories (0,74). On the other hand, the integral sustainability in mountain regions with natural restrictions is the lowest (0,56). These ecosystems' type has also the lowest (and close to the limits of satisfying level) levels for social sustainability, with the ecosystems in non-mountain regions with natural restrictions (0,52). Nevertheless, the ecological sustainability of agro-systems in mountain areas with natural restrictions is relatively high (0,58).

The integral sustainability of mountain ecosystems is on a medium level (0,58), but while its economic and social aspects are below the average for the country (respectively 0,61 and 0,53), the level of ecological sustainability is among the highest ones(0,6). The agricultural sustainability in the protected zones and territories is above the average for the country (0,62), these ecosystems having relatively high economic sustainability (0,74; the highest level of social sustainability (0,59) and good levels for ecological sustainability (0,58). the ecological sustainability in the plane-mountainous regions is the lowest in the country

(0,55), and for the non-mountainous regions with natural restrictions it is the highest (0,61).

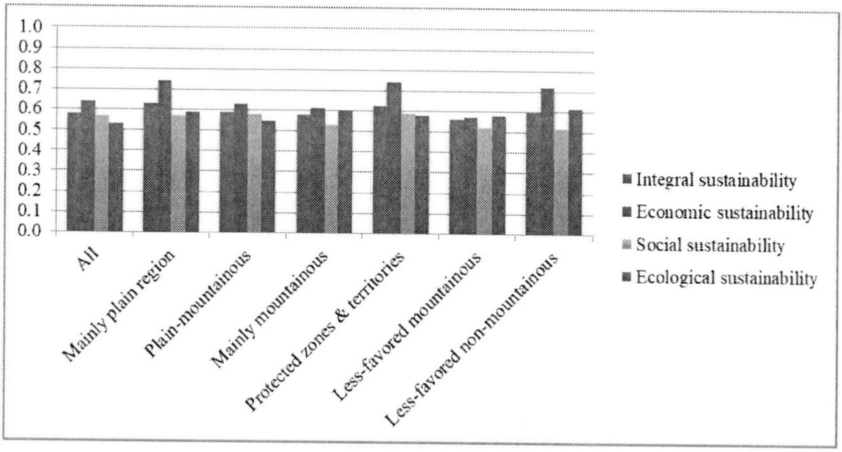

Source: Survey with managers of farms, 2017 and author's calculations.

Figure 5. Level of sustainability in the main types of agro-ecosystems in Bulgaria.

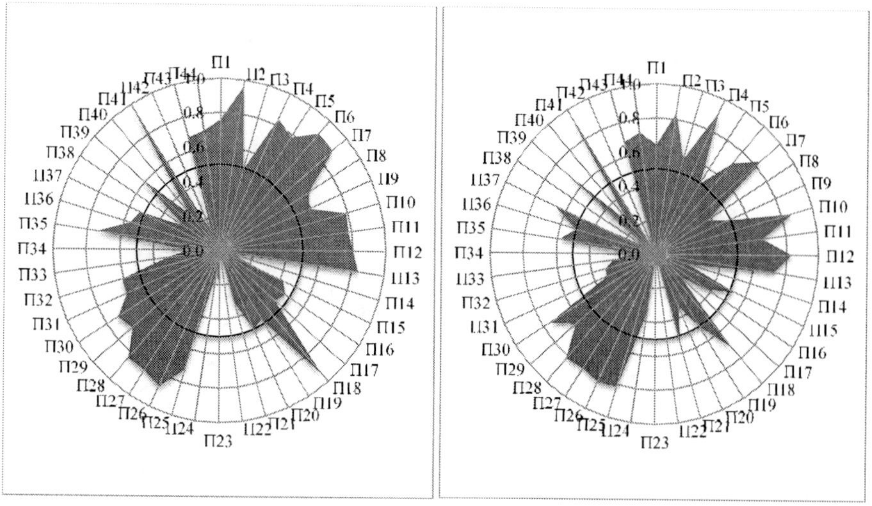

Predominantly mountainous regions **Protected zones and territories**

Figure 6. (Continued).

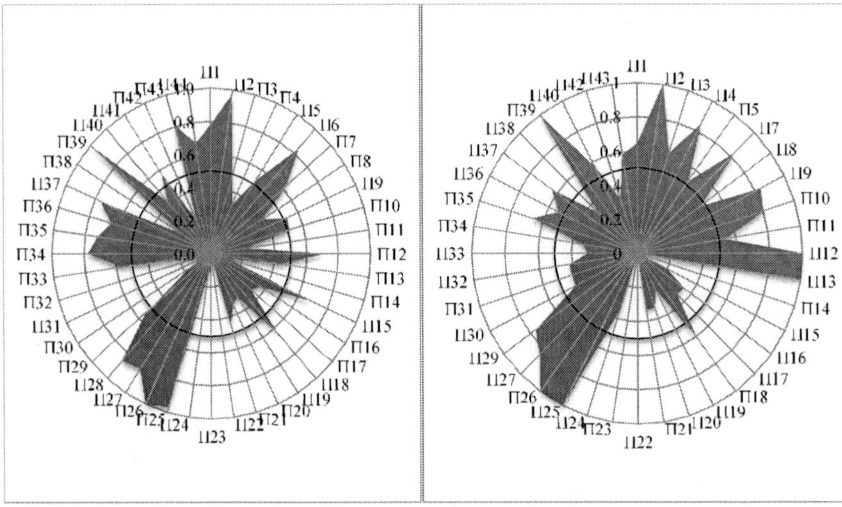

Mountainous with natural restrictions **Non-mountain with natural restrictions**

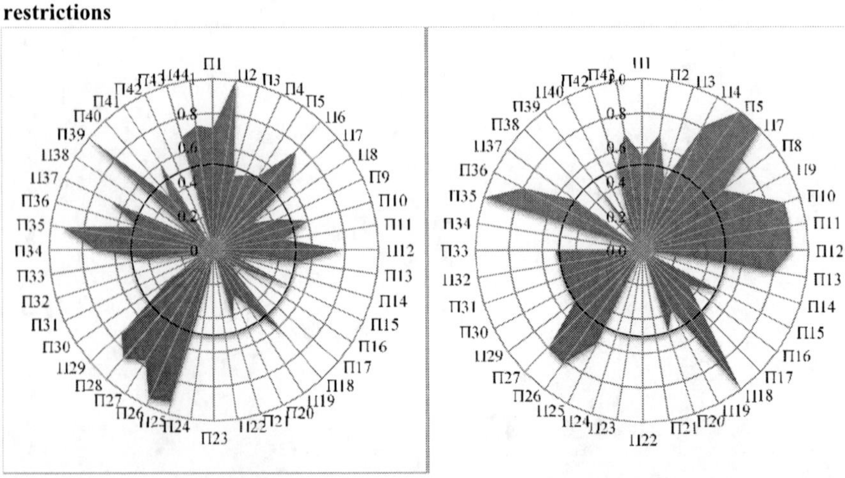

Source: Survey with managers of farms, 2017 and author's calculations.

Figure 6. Indicators for in the main agro-ecosystems types in Bulgaria.

The agriculture of ecosystems in the plane regions has high significances for economic sustainability for the indicators: share of own capital in the total capital (0,96), labour productivity (0,84), livestock productivity (0,9) and share of sold production in the total output (0,89) (Figure 6). The social sustainability of the sector in these regions is high in relation to degree of correspondence to the normative labour conditions

(0,84), education level of manager (0,94) and share of unoccupied seasonal labour positions in the total number of employed (0,87). Agriculture in such regions is with ecologically strong sustainability for the dynamics of UAA in the last 5 years (0,83), the dynamics of the raised livestock number in the last 5 years (0,83) and keeping the norms of animal welfare (1).

Simultaneously, the levels of some indicators in the plane agro-ecosystems have low levels. While the economic sustainability is satisfying only regarding the relation profit/production costs (0,49), for the social sustainability satisfying are the levels for number of family members working in the farm (0,42), manager's age (0,47), participation in training programs in the last 3 years (0,44), share of employed with special agricultural education/qualification (0,47) and number of participation in professional organizations and initiatives (0,31). Along with that, regarding the public position of the farmer, manager or owner (0,19) and participation in local initiatives (0,13) the state is unsatisfying and for presence of family member ready to take the farm (0,06), on the limit of the unsustainability. Moreover, according the indicator share of hired workers, members of trade unions, the state is unsustainability. The ecological sustainability of the sector in these regions is satisfying in relation to the share of arable land in the total agricultural land (0,32), presence of protected species on the farm territory (0,25) and number of cultural species (0,27); and unsatisfying for the keeping of practices for landscape maintenance (0,19) and implementation of principles for organic production (0,11).

In ecosystems of plane-mountain regions the economic sustainability of agriculture is high regarding the: share of own capital in the total (0,84), labour productivity (0,91) and share of sold production in the total output (0,84) (Figure 6). The highest in social aspect in these regions are the indicators: net farm income/average income in the region (0,87), degree of satisfaction from the farming activity (0,83), share of non-occupied permanent work positions in the total number of employed (0,81) and share of unoccupied seasonal work positions in the total number of employed (0,83). From ecological aspect, the best of these ecosystems are only the dynamics of the number of livestock in the last 5 years (0,82) and the keeping of norms of animal welfare (1).

At the same time agro-ecosystems in the plain-mountainous regions have satisfying values of economic sustainability for the growth of sales in the last 3 years (0,38) and investments growth in the last 5 years (0,49). The social sustainability in these regions is on satisfying levels in relation to manager's age (0,37), degree of participation of women in the farm management (0,33) and participation in local initiatives (0,33); unsatisfying regarding the presence of family member, ready to take the farm (0,2) and participation in training programs in last 3 years (0,2); and socially unstable for the share of hired workers, members of trade unions and public positions of the farmer, manager or owner. In the plane-mountain ecosystems the ecological sustainability is satisfying regarding the compliance with the norms of the fertilization with potassium (0,32), compliance with the norms of phosphorus fertilization (032) and share of arable land in the total agricultural land (0,26); unsatisfying for the keeping of practices for landscape maintenance (0,13), presence of protected species on the farm territory (0,07), and number of cultural species (0,24); and unstable for the implementation of principles for organic production.

The agricultural sustainability in ecosystems in mountain regions has the highest values for the economic indicators: share of own capital in the total capital (0,97)and livestock productivity (0,84); the social indicators of the share of non-occupied permanent work positions in the total number of employed (0,97), and share of unoccupied seasonal work positions in the total number of employed (1); and ecological indicators: dynamics of UAA in last 5 years (0,83), dynamics of raised livestock in last 5 years (0,86), natural biodiversity protection (1), and yield variation of the main crops for 5 years (0,81) (Figure 6). In mountain regions with satisfying values for sustainability are the economic relation profit/production costs (0,49), labour productivity (0,33), and sales' growth in last 3 years (0,38). The social sustainability of this type of ecosystems is satisfying in lots of indicators: degree of compliance with normative labour conditions (0,44), manager's age (0,37), participation in training programs in last 3 years (0,33), share of employed with special agricultural education/qualification

(0,31), degree of participation of women in the farm management (0,33), and number of participations in professional organizations and initiatives (0,44). Furthermore, the social sustainability is unsatisfying in relation to the payment of hired labour/average income in the region (0,22), presence of a family member, ready to take the farm (0,11), public position of the farmer, manager or owner (0,11), and participation in local initiatives (0,11). In relation to the share of hired workers, members of trade unions, there is a social instability. In the mountain agro-ecosystems the ecological sustainability is on a satisfying level for the number of cultural species (0,41), and unsatisfying for the compliance with the norms of nitrate fertilization (0,17), compliance with the norms for potassium fertilization (0,08), compliance of phosphorus fertilization with the norms (0,08), presence of protected species on the farm territory (0,22), and implementation of principles for organic production (0,22).

The ecosystems' agricultural sustainability in the protected zones and territories is economically high regarding the share of own capital in the total one (1), labour productivity (0,85), share of sold production in the total output (0,83), and investments' growth in the last 5 years (0,84) (Figure 6). This ecosystem type has strong social stability for the degree of satisfaction of the farming activity (1), degree of compliance with the normative labour conditions (1), share of unoccupied permanent work positions in the total number of employed (1), and share of non-occupied seasonal work positions in the total number of employed (1). In ecological aspect the agricultural sustainability in the protected zones and territories is high only regarding the dynamic of UAA in last 5 years (0,83), and natural biodiversity protection (1). On the other hand, the economic sustainability of agro-ecosystems with protected zones and territories is satisfying for the sales' growth in the last 3 years (0,47), while for the livestock productivity there is an instability. The social sustainability in these zones and territories is on satisfying level in relation to manager's age (035), participations in training programs in last 3 years (0,33), degree of participation of women in the farm management (0,33), number of participations in professional organizations and initiatives (0,33), and participation in local initiatives (0,33). For the social indicators the number of family members working in the farm (0,2), and share of employed with special agricultural education/qualification (0,24) the sustainability level is unsatisfying. Moreover, regarding the presence of a family member ready to take the farm, the share of hired workers, members in trade union and the public position of the farmer, manager or owner, the ecosystems are unsustainable. In protected zones and territories some ecological indicators are also relatively low (unsatisfying): compliance to norms of the fertilization with potassium (0,42), compliance to norms of the fertilization with phosphorus (0,42), share of arable land in the total agricultural land (0,3), keeping of practices for landscape maintenance (0,33), presence of protected species on the farm territory (0,33) and implementation of principles for organic production (0,33).

Agricultural sustainability in ecosystems of mountain regions with natural restrictions are highly economically sustainable just in relation to the share of own capital in the total (1); strongly socially sustainable for the share of unoccupied permanent work positions in the total number of employed (0,93) and the share of unoccupied seasonal work positions in the total number of employed (0,96); and highly ecologically sustainable according the dynamics of the livestock number in last 5 years (0,84), degree of pollution of underground waters with nitrates (0,93) and protection of natural biodiversity (1) (Figure 6). At the same time, some economic indicators of sustainability in these ecosystems are on satisfying level, as: profit/production costs (0,45), labour productivity (0,48), sales' growth in last 3 years (0,29), and investments' growth in last 5 years (0,43). Similarly, the social sustainability of this ecosystems' type is satisfying regarding: payment of hired labour/average income in the region (0,43), share of employed with special agricultural education/qualification (0,38), degree of participation of women in the farm management (0,29) and number of participations in professional organizations and initiatives (0,43). The level of social sustainability in such regions is unsatisfying for presence of family member, ready to take the farm (0,14), manager's age (0,19), participation in training programs in last 3 years (0,14) and participation in local initiatives (0,14). In relation to the share of hired workers, members of trade unions and public position of manager, farmer and owner, the mountain regions with natural restrictions are socially unsustainable. In these regions some indicators for ecological sustainability have satisfying levels, as the compliance to norms of the nitrate fertilization (0,32), share of arable land in the total agricultural land (0,4), level of fuel consumption (0,49) and number of cultural species (0,4). The ecological sustainability is unsatisfying for the compliance to the norms of potassium fertilization (0,11), compliance to norms of phosphorus fertilization (0,11) and presence of protected species on the farm territory (0,14), while for the principles of organic production implementation, they are unsustainable.

The agricultural sustainability in the non-mountain regions with natural restrictions is economically high regarding the labour productivity (0,81), land productivity (1) and share of sold output in the total one (1) (Figure 6). In relation to the social sustainability, the indicators are high for: net farm income/average income in the region (0,9), payment of hired work in the region (0,9), degree of satisfaction from the farming activity (0,9), education level of manager (1) and share of unoccupied seasonal work positions in the total number of employed (0,81). The ecological sustainability in these regions is high only for the pollution of underground waters with nitrates (1). The agro-ecosystems in the non-mountain regions with natural restrictions have satisfying economic sustainability only regarding the ratio profit/production costs (0,43). The social sustainability of these agro-ecosystems is satisfying for the age of manager (0,34) and share of employed with special agricultural education/ qualification (0,38). As regards to the presence of a family member ready to take the farm; number of participation in professional organizations and initiatives; share of hired workers, members of trade unions; public position of farmer, manager or owner and participation in local initiatives, these ecosystems are unsustainable. Non-mountain regions with natural restrictions have unsatisfying level of ecological sustainability for the indicator number of cultural species (0,15) and they are ecologically unsustainable as regards the keeping of landscape maintenance practices (0) and presence of protected species on the farm territory. (0).

Level of Agricultural Sustainability in the Specific Agro-Ecosystems

In the fourth geographical regions of the country have been identified and analyzed the following agro-ecosystems, important for the respective region and for the country, as a whole: the ecosystems alongside the rivers Yantra, Maritsa and Struma, West Thrace valley, Middle Danube plane, Doupnitsa and Sandanski-Petrich valleys, South- cost Black sea, Sashtinska Sredna Gora mountain and West Rila mountain.

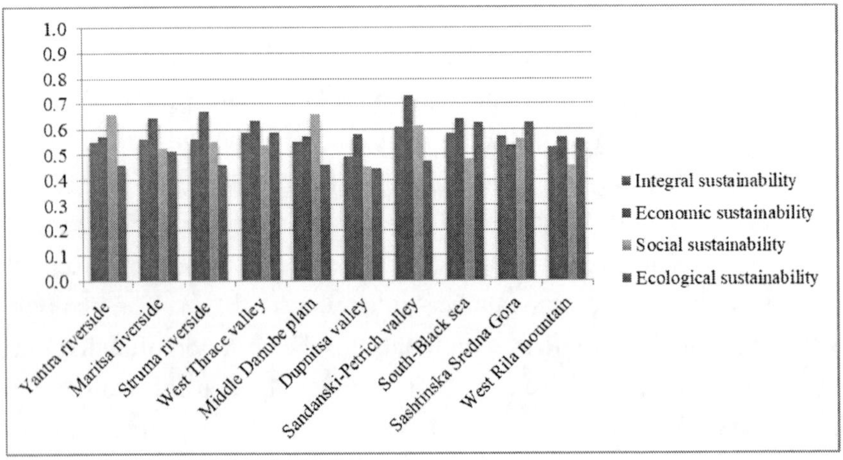

Source: Survey with managers of farms, 2017 and author's calculations.

Figure 7. Levels of sustainability in the specific agro-ecosystems in Bulgaria.

The assessment postulated that there is a big variation in the levels of integral, economic, social and ecological sustainability of agriculture in the specific ecosystems. From the analyzed 10 agro-ecosystems, the highest integral sustainability has Sandanski-Petrich valley (0,61) with highest values of economic sustainability (0,73), social sustainability with also high values (0,61), while the ecological sustainability is among the lowest in the country and on satisfying level (0,47) (Figure 7). On the other hand, the integral sustainability of agriculture in Dupnitsa valley is on the lowest level (0,49) and the only one with satisfying level among the analyzed ecosystems. In this ecosystems the levels of social (0,45) and ecological (0,45) sustainability are satisfying and the lowest among the analyzed.

The integral sustainability of agro-ecosystems in the areas alongside the rivers Yantra, Maritsa and Struma is on a relatively low (under the average) level – respectively 0,55, 0,56 и 0,56. However, there is a big differentiation of different aspects of sustainability in these specific ecosystems. For the eco-system alongside Struma river the economic sustainability is on a high level (0,67), while for Yantra riverside it is slightly below the average for the country. On the other hand, the area alongside Yantra has the highest level of social sustainability (0,66), whereas the area alongside Maritsa has the lowest social sustainability and close to the limit of the satisfying level (0,52). For the three riverside ecosystems the ecological sustainability of the sector is below the average values for the country, as for Maritsa riverside the value is on the border of the satisfying level (0,51), and for the other riverside ecosystems – on satisfying level (by 0,46).

The agro-ecosystem Middle Danube plain has relatively low integral sustainability (0,55), with levels of social sustainability among the highest in the country (0,66), and from ecological aspect on a satisfying level (0,46) and among the lowest for the country. The agriculture in the West Thrace valley has integral sustainability on a relatively high level and over the average for the country (0,59). This agro-ecosystem has good economic sustainability, over the average (0,67), with one of the highest levels of ecological sustainability (0,59), but relatively low and under the average social sustainability (0,54).

Both analyzed specific mountain agro-ecosystems have lower integral sustainability than the average – respectively 0,57 for Sashtinska Sredna Gora mountain, and 0,53 for West Rila mountain. The social (0,56) and the ecological (0,63) sustainability of Sashtinska Sredna Gora are higher than the values of West Rila mountain (respectively on satisfying level 0,46 and good level 0,56), whereas for the economic sustainability is the opposite (0,53 and 0,57). Sashtinska Sredna Gora and South Black sea cost have the highest indicators for ecological sustainability among all analyzed specific ecosystems in the country. The integral sustainability of agriculture of

South Black sea is on the average level for the country - 0,58, while the economic sustainability is on a middle level (0,64), the social sustainability is satisfying (0,48), and the ecological is the best of all analyzed (0,63).

Yantra riverside **Maritsa riverside**

Struma riverside **South-Black sea**

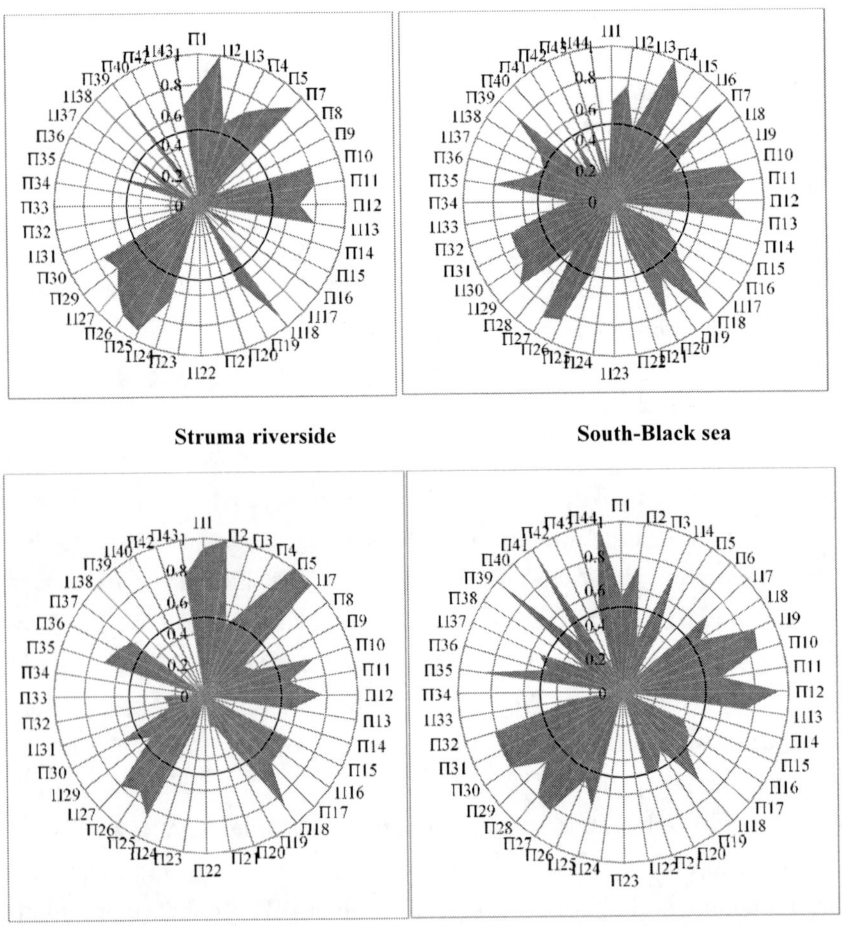

Middle Danube plain **West Thrace valley**

Figure 8. (Continued).

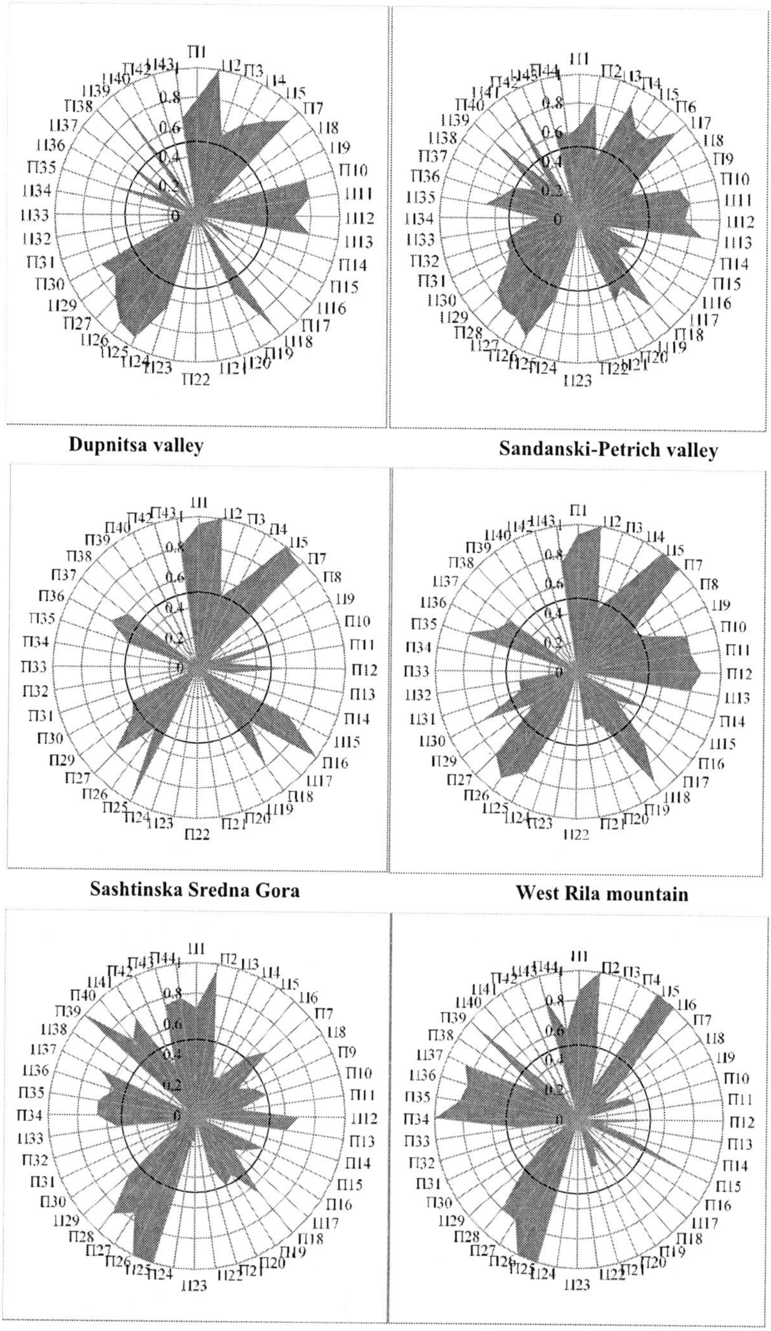

Source: survey with managers of farms, 2017 and author's calculations.

Figure 8. Indicators for sustainability in the specific agro-ecosystems in Bulgaria.

There is a considerable variation of different indicators' levels in the specific agro-ecosystems. Three specific riverside ecosystems in North Central, South Central and South-West regions were analyzed. In the agro-ecosystem of Yantra river high levels have only the indicators for economic sustainability – share of own capital in the total one (1) and share of sold production in the total output (0,91); the indicators for social sustainability – level of education of the manager (0,93), number of participations in professional organizations and initiatives (1), share of unoccupied permanent work positions in the total number of employed (0,93), and share of unoccupied seasonal work positions in the total number of employed (0,9); and for the ecological sustainability – natural biodiversity protection (1) (Figure 8).

The agriculture of Yantra riverside has unsatisfying sustainability for lots of indicators: economic growth of sales in the last 3 years (0,13) and investments' growth in the last 5 years (0,2); social number of family members, working in the farm (0,2); and ecological: compliance of potassium fertilization to the norms (0,17), compliance to the norms of phosphorus fertilization (0,17), level of fuel consumption (0,25) and number of cultural species (0,2). Moreover, this system is unsustainable due to lots of social and ecological indicators: presence of a family member, ready to take the farm; participation in training programs in last 3 years; degree of participation of women in the farm management, share of hired workers, members of trade unions; public position, occupied by the farmer, manager or owner; share of arable land in the total agricultural land; keeping of practices for landscape maintenance; presence of protected species on the farm territory; implementation of principles for organic production. In relation to the age of manager, the social sustainability is satisfying (0,32). Similar to indicators of the agro-ecosystem along Yantra riverside are the indicators for the sustainability of Middle Danube plain.

The agriculture in the other analyzed riverside ecosystem, of Maritsa, is characterized by several indicators for levels of high sustainability: economic – labour productivity (1), land productivity (0,81) and share of sold production in the total production (0,98); social – payment of hired labour/average income in the region (0,88), degree of compliance to normative labour conditions (0,88), education level of the manager (0,97), degree of participation of women in the farm management (0,86), share of unoccupied seasonal work positions in the total number of employed (0,84); and ecological – dynamics of UAA in the last 5 years (0,88), soil erosion (0,83), degree of pollution of underground waters with nitrates (0,81) and natural biodiversity protection (0,86) (Figure 8).

The agro-ecosystems from the riverside of Maritsa have satisfying sustainability of economic indicators: profit/ production costs (0,48), livestock productivity (0,4) and investments' growth in the last 5 years (0,43). The level of social indicators is also satisfying: number of family members, working in the farm (0,36), manager's age (0,48), number of participations in professional organizations and initiatives (0,29) and share of unoccupied permanent work positions in the total number of employed (0,44). Similar is the level of ecological indicators: dynamics of the arable land in the last 5 years (0,4) and share of arable land in the total agricultural land (0,44). The agricultural sustainability alongside Maritsa river is on unsatisfying level about the social and ecological indicators: participation in local initiatives (0,14), keeping of practices for landscape maintenance (0,29), number of cultural species (0,24), implementation of principles for organic production (0,14) and percentage of mortality of the livestock for 5 years (0,2). In relation to social dimensions there is a state of unsustainability: presence of family member ready to take the farm, share of hired workers, members in professional organizations and public position of the farmer, manager or owner.

Unlikely the other two riverside agro-ecosystems, this of Struma river has high economic levels of sustainability for the share of direct payments in the net income (0,94), share of own capital in the total one (1), land productivity (1) and share of sold production in the total output (0,99) (Figure 16). The social sustainability in this agro-ecosystem is high only regarding the education level of the manager (0,88) and share of unoccupied work positions in the total number of employed (0,86). On the other hand, some indicators of economic sustainability in this agro-ecosystem have satisfying levels, as: profit/production costs (0,47), growth of sales in the last 3 years (0,32) and investments' growth in the last 5 years (0,36). Similar is the level of sustainability regarding the social and ecological indicators for the employed with special agricultural education/qualification (0,34), soil erosion (0,44) and share of arable land in the total agricultural land (0,28).

Moreover, theagro-ecosystem of Struma riverside is unsustainable in relation to the social measurers: degree of participation of women in the farm management (0,2), number of participation in professional organizations and initiatives (0,2) and participation in local initiatives (0,2); and ecological indicators: compliance to the norms of potassium fertilization (0,25), compliance to the norms of phosphorus fertilization (0,25) and number of cultural species (0,12). This agro-ecosystem is socially unsustainable in relation to the participation of a family member, ready to take the farm; share of hired workers, members in trade unions and public position of the farmer, manager or owner. The ecosystem is also in state of ecological unsustainability regarding the keeping of practices for landscape maintenance, presence of protected species on the farm territory, protection of the natural biodiversity and implementation of principles of organic production.

The agricultural sustainability in the South-Black sea ecosystem has high levels for the economic indicator - investments' growth in the last 5 years (0,88) and for the social indicators: net farm income/average income in the region (0,85) and degree of satisfaction from farming activity (0,95) (Figure 10). The agro-ecosystem is also ecologically sustainable with lots of indicators: dynamics of UAA in the last 5 years (0,82), compliance to the norms of nitrate fertilization (0,81), compliance to the norms of the potassium fertilization (0,81), compliance to the norms of the phosphorus fertilization (0,81), degree of pollution of underground waters with nitrates (0,87), natural biodiversity protection (1), keeping the norms of animal welfare (1) and percentage of livestock mortality for 5 years (1). The agro-ecosystem South-Black sea has satisfying sustainability concerning the economic indicator profit/production costs (0,31); several social indicators, as: number of family members working in the farm (0,4), manager's age (0,47) and share of employed with special agricultural education/

qualification (0,47); and ecological indicators for : share of arable land in total agricultural land (0,31), level of fuel consumption (0,47) and number of cultural species (0,37).

This specific ecosystem has unsatisfying sustainability of agriculture regarding the economic aspect for livestock productivity (0,11) and from ecological aspect: for the presence of protected species on the farm territory (0,25) and implementation of organic production principles (0,12). The agriculture of South-Black sea is socially unsustainable regarding the presence of a family member ready to take the farm; share of workers, members of trade unions; public position of the farmer, manager or owner and participation in local initiatives, and in ecological aspect, for the keeping of practices for landscape maintenance.

The agriculture in the West Thrace valley has high economic sustainability regarding the indicators: share of own capital in the total one (0,82), labour productivity (0,88) and share of sold production in the total (0,92); high social sustainability for compliance to the normative labour conditions (0,89) and share of unoccupied seasonal work places in the total number of employed (0,89); and high ecological sustainability for the dynamics of UAA in the last 5 years (0,82), dynamics of the livestock number in the last 5 years (0,82), natural biodiversity protection (0,82), and keeping of norms for animal welfare (1) (Figure 10). The agriculture of this ecosystem has satisfying levels of economic sustainability for: profit/ production costs (0,44)and investments' growth in the last 5 years (0,4); social sustainability for: number of family members working in the farm (0,48), manager's age (0,36), participation in training programs in last 3 years (0,36); and ecological sustainability for: share of arable land in the total agricultural land (0,4), keeping of practices for landscape maintenance (0,27), presence of protected species on the farm territory (0,36) and number of cultural species (0,3).

The social sustainability is unsatisfying for indicators: presence of family member ready to take the farm (0,18), number of participations in professional organizations and initiatives (0,18) and participation in local initiatives (0,18), and regarding the share of hired, members of trade unions, and public position of farmer, manager or owner the state is unsustainable. The same state has the ecological sustainability regarding the implementation of principles for organic production (0,09).

In the South-West region of the country have been analyzed two specific agro-ecosystems of Dupnitsa valley and of Sandanski-Petrich valley. Dupnitsa valley has high economic sustainability of indicators: share of direct payments in the net income (0,95), share of own capital in the total one (1), land productivity (1) and share of sold output in the total (0,97) (Figure 16). The agriculture in this ecosystem has high social and ecological sustainability only regarding the age of the manager (1), share of unoccupied permanent work positions in the total number of employed (1) and variation of yields of the main crops for 5 years (0,81).

Under two economic, several social and one ecological indicator, the sustainability of this agro-ecosystem is unsatisfying: sales growth in last 3 years (0,1), investments' growth in last 5 years (0,1), payment of hired labour/average income in the region (0,2), degree of compliance to normative labour conditions (0,22), and share of employed with specific agricultural education/qualification (0,2), and number of cultural species (0,1). Under many social and ecological indicators the level is unsustainable: presence of a family member ready to take the farm; degree of participation of women in the farm management; number of participations in professional organizations and initiatives; share of hired workers, members of trade unions; public position of the farmer, manager or owner; participation in local initiatives; compliance to the norms of potassium fertilization; compliance to the norms of phosphorus fertilization; respecting of practices for the landscape maintenance; presence of protected species on the farm territory; protection of natural biodiversity and implementation of organic production principles.

Other analyzed agro-ecosystem is Sandanski-Petrich valley, which is characterized by high sustainability of economic indicators: share of direct payments in the net income (0,93), share of own capital in the total (1), land productivity (1) and share of sold output in the total output (1); social measurers: degree of satisfaction from farm activity (0,86), education level of manager (0,93) and share of unoccupied seasonal work positions in the total number of employed (0,9); and ecological indicator: degree of pollution of underground waters with nitrates (0,83). In this ecosystem the agricultural sustainability has relatively low (satisfying) economic sustainability according two indicators: profit/production costs (0,45) and growth of sales in the last 3 years (0,47). Similarly, the social sustainability in the agro-ecosystem has satisfying levels in relation to: manager's age (0,33); share of employed with special agricultural education/qualification (0,44); degree of participation of women in the farm management (0,33); number of participation in professional organizations and initiatives (0,33) and participation in local initiatives (0,33). The agriculture in this area is socially unsustainable regarding the presence of a family member, ready to take the farm; share of hired workers, members of trade unions and public position of the farmer, manager or owner.

Apart this, the ecological sustainability of Sandanski-Petrich valley is satisfying for the soil erosion (0,37); compliance to norms of potassium fertilization (0,42) and compliance to norms of phosphorus fertilization (0,42); unsatisfying regarding the share of arable land in the total agricultural land (0,1) and number of cultural species (0,13); and ecologically unsustainable regarding the keeping of practices for landscape maintenance; presence of protected species on the farm territory; protection of natural biodiversity and implementation of organic production principles.

Two mountain agro-ecosystems have been analyzed – Sashtinska Sredna Gora and Western Rila mountain. The agriculture in Sashtinska Sredna Gora is economically sustainable regarding the share of own capital in the total (0,96); strongly socially sustainable for the share of unoccupied permanent work positions in the total number of employed (1) and share of unoccupied seasonal work positions in the total number of employed (1); and highly ecologically sustainable for the dynamics of the livestock number in the last 5 years (0,85) and for the natural biodiversity protection (1) (Figure 8). The agricultural production in this ecosystem has satisfying levels of many economic and social indicators: profit/production costs (0,43), labour productivity (0,27), land productivity (0,3), sales growth in last 3 years (0,33), investments growth in last 5 years (0,43), payment of hired labour/average income in the region (0,3), manager's age (0,41), participation in education programs in last 3 years (0,33), share of employed with special agricultural education/qualification (0,45) and number of participations in professional organizations and initiatives (0,33). This agro-ecosystem has satisfying ecological sustainability in relation to the implementation of organic production principles (0,33).

Moreover, according several social and ecological indicators, the agriculture in Sashtinska Sredna Gora is with unsatisfying sustainability: public position of the farmer, manager or owner (0,17), participation in local initiatives (0,17), compliance to norms of the nitrate fertilization (0,17), compliance to norms of the potassium fertilization (0,12), compliance to norms of the phosphorus fertilization (0,12). This agro-ecosystem is socially and ecologically unsustainable in relation to the presence of a family member, ready to take the farm; share of hired workers, members of trade unions and presence of protected species on the farm territory.

The other mountain agro-ecosystem Western Rila mountain has high economic sustainability in relation to the share of direct payments in the net income (0,87), share of own capital in the total (1), land productivity (1) and livestock productivity (1) (Figure 8). The social sustainability is strong regarding the indicators: number of family members working in the farm (0,86), share of unoccupied permanent work positions in the total number of employed (1) and share of unoccupied seasonal work positions in the total number of employed (1). The agriculture in Western Rils mountain is ecologically sustainable for the respecting of practices for landscape maintenance (1), degree of pollution of underground waters with nitrates (0,83), level of consumption of electricity (0,87), protection of natural biodiversity (1) and variation of yields of main crops for 5 years (0,83). This agro-ecosystem has satisfying economic sustainability in relation to profit/production costs (0,43), share of sold output in the total output (0,41) and investments growth in last 5 years (0,37). The level of social sustainability is satisfying for the net farm income/average income in the region (0,4), presence of a family member, ready to take the farm (0,33), degree of participation of women in the farm management (0,33) and number of participation in professional organizations and initiatives

(0,33). The agricultural sustainability is unsatisfying regarding the economic indicators labour productivity (0,22) and sales growth in the last 3 years (0,2); and social indicators degree of compliance to normative labour conditions (0,15) and share of employed with special agricultural education/qualification (0,2). Furthermore, some social indicators in this agro-ecosystem have unsustainability levels: payment of hired labour/average income in the region, manager's age, participation in education programs in the last 3 years, share of hired workers, members in trade unions, public positions of the farmer, manager or owner, participation in local initiatives.

The agro-ecosystem Western Rila mountain has satisfying ecological sustainability for: soil erosion (0,46), share of arable land in the total agricultural land (0,42), presence of protected species on the farm territory (0,33) and respecting the norms for animal welfare (0,33). The ecological sustainability of the ecosystem is unsatisfying for: compliance to norms of nitrate fertilization (0,25), number of cultural species (0,23), compliance to norms of potassium fertilization (0,08) and compliance to norms of phosphorus fertilization (0,08). This ecosystem is ecologically unsustainable in relation to the principles of organic production.

Sustainability Contribution of Different Sub-Sectors of Agriculture and Type of Farms

Our analysis allows to assess the contribution of different sub-sectors and farms with different specialization to the total agricultural sustainability and its main aspects. The highest integral sustainability has been shown by the mixed livestock-breeding (0,7) and mixed crop-growing (0,66) farms, followed by the perennial crops farms (0,63). (Figure 9). Therefore, the mixed livestock-breeding and crop-growing farms and the farms with perennials contribute in highest degree for improving the integral sustainability of Bulgarian agriculture. From the other hand, the farms specialized in pigs, poultry and rabbits (0,53); vegetables, flowers and mushrooms (0,54) and mixed livestock-crops (0,54) have the lowest integral sustainability. This means that the last mentioned types of farms decrease in a biggest degree the integral sustainability in the country.

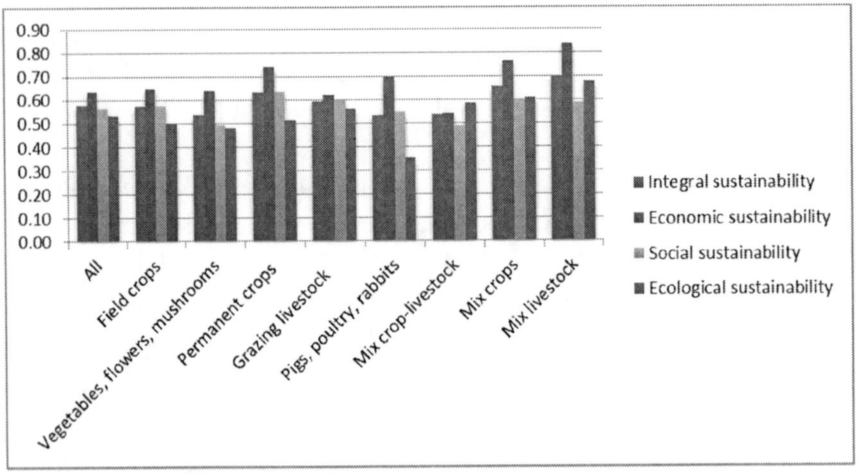

Source: Survey with managers of farms, 2017 and author's calculations.

Figure 9. Sustainability contribution of different sub-sectors of agriculture in Bulgaria.

Similar to integral sustainability, the sub-sectors with the highest economic sustainability are: mixed livestock breeding (0,84), mixed crop growing (0,76) and perennial crops (0,74). The mixed crop-growing production has the highest ecological sustainability (0,61) and one of the best social sustainability (0,6). The perennial crops sector has high social sustainability (0,64), but lower than the average and almost satisfying ecological sustainability (0,51). The social sustainability of farms specialized in grazing livestock has comparatively high level of social sustainability (0,6). The social sustainability in mixed crop-livestock farms has a satisfying level (0,49). The pigs, poultry and rabbits' farms have lowest and satisfying level (0,35), like the farms for vegetables, flowers and mushrooms (0,48). The field crops farms have good, but relatively low ecological sustainability (0,5), close to the satisfying level.

Furthermore, the different agricultural sub-sectors are characterized by important variation of levels of sustainability indicators and therefore, different type of contribution to overall and aspect level of sustainability of agri-ecosystems in the country.

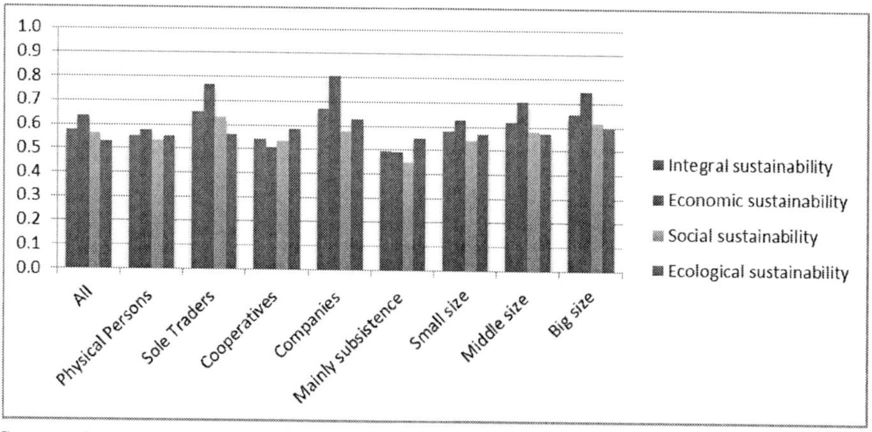

Source: Survey with managers of farms, 2017 and author's calculations.

Figure 10. Sustainability contribution of farms of different types in Bulgaria.

Similarly, the agricultural sustainability in different farm types has various levels, which is determined by the specific contribution of different farms for the formation of the existing level of sustainability in the agri-ecosystems of the country.

Among the farms with different juridical status, the trade associations show the highest agricultural sustainability (0,67), contribute the most for the agricultural sustainability of the country. In these organizational and management structures the economic (0,8) and ecological (0,63) aspects of agricultural sustainability have the highest levels, while the social sustainability is on average for the country level (Figure 10). The social sustainability is highest for sole traders (0,63), whose integral (0,65) and economic (0,77) sustainability is on the second place and are close to the values of the trade associations.

The agricultural production in cooperatives has the lowest integral sustainability (0,54), which economic sustainability (0,51) is on the border with the satisfying level, and the social sustainability is the lowest, the same level as for individuals (0,53). The cooperatives have ecological sustainability of the production on relatively high level (0,59). The agricultural production of individuals has integral sustainability under the average level (0,55) with lower than the average for the economic (0,58) and social (0,53) sustainability.

The agricultural sustainability in farms with different market orientation and sizes is also characterized by different levels and contribution to the integral agricultural sustainability in the country (Figure 10). The highest integral sustainability is shown by the large farms (0,65), having the highest economic (0,75), social (0,62) and ecological (0,6) sustainability. Therefore, these farms contribute in biggest degree for the increase of the integral level of agricultural sustainability in the country. In predominantly self-subsistence farms the agricultural sustainability is low, close to the satisfying level (0,5). In these farms all the aspects of the agricultural sustainability have low levels, in comparison to the large and market oriented farms, as the economic (0,49) and social (0,45) sustainability are satisfying. There is a trend to decrease of the levels of integral, economic and social sustainability with the decrease of the farm sizes. The ecological sustainability of farms with small and medium sizes has the same levels, which are lower than these of the bigger farms, but higher than the levels of self-subsistence farms.

The individual indicators for the sustainability of farms of different juridical kind, size and market orientation are very differentiated, demonstrating different type of contribution of overall and aspect sustainability of respected agro-ecosystems.

Comparison of Assessment of Agricultural Sustainability with the Previous Studies in the Area

Finally, we compare the integral agricultural sustainability based on the assessment of sustainability of agro-ecosystems with the results of previous studies assessing agrarian sustainability with the aggregate sectoral (statistical, etc.) data in Bulgaria (Bachev et al., 2017).

According to the previous study based on aggregate data, using the same methodological approach, the integral sustainability index of the Bulgarian agriculture is 0.58, which corresponds to a Good sustainability.

That study has found out that the Economic sustainability of the Bulgarian agriculture is Good (index of sustainability 0.7), while the Social and the Environmental sustainability are also as Good but with a lower index (for both of them it is 0.53), close to satisfactory level.

Therefore, the integral assessment results based on the micro agro-ecosystems (farm) data are similar with the results based on aggregated sectoral (statistical, etc.) data. It means that both approaches are reliable and could be simultaneously used for assessing agricultural sustainability at various level – sector, subsector, region, agro-ecosystem, and farm.

Factors for Improving the Sustainability of Agro-Ecosystems in Bulgaria

Diverse social, economic, market-related, ideological, and personal factors stimulate or restrict the activities of farming in terms of sustainable operation and development.

According to the managers surveyed, the factors encouraging farming enterprises to improve their economic sustainability include: market demand and price; direct state subsidies; market competition; financial capability; participation in public support programs; possibility of benefitting immediately; possibility of benefitting in the near future; tax preferences; possibility of benefitting in the long term; and integration with buyers of farm products. Factors considered critical by a smaller proportion of enterprises include: regional community initiatives and pressure; social recognition of individual contribution; pressure and initiatives of interest groups; immediate benefits for other people and groups; and professional training for managers and hired labor.

The factors encouraging the enhancement of social sustainability for the greatest number of farms include: personal convictions and satisfaction; social recognition of individual contribution; immediate benefits for other people and groups; regional community initiatives and pressure; access to advisory services; European Union policy; and existing regional problems and risks. For a small number of enterprises, the important factors encouraging the social sustainability include: state control and sanctions; existence of long-term contracts with the state; registration and certification of products and services; tax preferences; and integration with suppliers.

The factors encouraging the environmental sustainability include: problems and risks existing at the global scale; official regulations, standards, and norms; existing regional problems and risks; and European Union policies. Significant factors encouraging the ecological sustainability for a small number of enterprises include: integration with suppliers; tax preferences; existence of long-term contracts with the state; market demand and price; integration with buyers; market competition; initiatives and pressure from interest groups; partners available for cooperative activities; initiatives of other farmers; and the possibility of garnering immediate benefits.

These motives need to be examined in relation to the modernization of public policy and the establishment of programs for sustainable development of agro-ecosystems in Bulgaria.

This survey has found that current public policies and diverse instruments of public support that improve the economic sustainability of farming enterprises in Bulgaria include: direct area-based payments; national top-ups for products and livestock; modernization of agricultural holdings; green payments; support for semi-market farms. Measures that could considerably improve the economic sustainability of a small number of holdings include: afforestation and restoration of forest; restoration and development of residential areas; stimulation of rural tourism; and the provision of services to residents of rural areas.

The impact that national and European policies have on the social and environmental sustainability of Bulgarian farming enterprises is relatively weak. Instruments that could augment the social sustainability of the majority of farming enterprises include: strategies for local development; the provision of services to residents of rural areas; restoration and development of residential areas; and stimulation of rural tourism. The social sustainability of a small number of holdings could be improved by ecological measures such as: payments for Natura 2000; agricultural environmental payments; and greater support for organic farming.

The most important actions to improve the environmental sustainability of farming enterprises include: green payments; support for organic farming; obligatory standards, norms, rules, and restrictions; and agro-environmental payments. Public instruments that would have the least impact on ecological sustainability of Bulgarian farming enterprises at the current stage of development include: support for setting up micro-enterprises; establishing produce organizations; support for semi-market farms; diversification into non-agricultural activities; support for young farmers; and restoration and development of residential areas

There is a difference shown between individual instruments of public policy and their impact on the sustainability of farming enterprises of different types and agro-eco-systems. Mechanisms and instruments of national and European policy with the greatest impact in improving the sustainability of Bulgarian farming enterprises include: 1) Obligatory standards, norms, rules, and restrictions in terms of the governance of big enterprises and the environmental sustainability of enterprises specializing in pigs, poultry, and rabbits. 2) Direct area-based payments to improve the economic sustainability of: sole traders, cooperatives, companies, holdings of small size for their sector; enterprises specializing in pigs, poultry, and rabbits, mixed crops, and permanent crops; and enterprises located in non-mountainous regions with natural handicaps, those with land in protected zones and territories, the majority of those in mountainous regions, mountainous regions with natural handicaps, and those in the southwest and south-central regions of the country. 3) National top-ups for products and livestock to improve the economic sustainability of: companies, holdings predominantly for subsistence, and those specializing in grazing livestock; the majority of those in mountainous regions, those with land in protected zones and territories, and those located in the north-central and southwest regions of the country; 4) Green payments to improve the economic sustainability of enterprises located in mountainous regions, those with land in protected zones and territories, and those in the southwest region of the country. 5) Professional training and advice for large enterprises. 6) The modernization of agricultural holdings to improve the economic sustainability of: sole traders and companies; those specializing in mixed livestock and mixed crops; and those located in mountainous regions and in the north-central and south-central regions. 7) Support for semi-market farms and the establishment of produce organizations to improve the economic sustainability of holdings predominantly for subsistence. 8) Natural handicap payments to farmers in mountainous areas to improve the economic sustainability of farming enterprises located in such areas.

All these data on the real impact that individual mechanisms and instruments of public support have on different aspects of sustainability among Bulgarian farming enterprises need to be taken into account when seeking to improve policies and programs supporting agricultural sectors and enterprises of diverse types and agro-ecosystems.

CONCLUSION

This first in its kind assessment on sustainability of agro-ecosystems in Bulgaria let make some important conclusions about the state of their sustainability, and recommendations for improvement of managerial and assessment practices.

The elaborated and experimented holistic framework gives a possibility to improve the general and aspects' sustainability assessment. That innovative approach has to be further discussed, experimented, improved and adapted to the specific conditions and evolution of agro-ecosystems of different types, as well as to the needs of decision-makers at various levels – farmers, interests groups, government officials, policy-makers, etc.

There is a considerable differentiation in the level of integral and aspects' sustainability in agricultural ecosystems of analyzed main and specific types. With the highest integral sustainability are the agro-ecosystems of the plane regions and Sandanski-Petrich hollow, while the least sustainable are agro-ecosystems of mountain regions with natural restrictions and Dupnitsa hollow. Furthermore, there are substantial variations in the levels of the economic, social and ecological sustainability of agro-ecosystems of different type. What is more, the individual indicators with the highest and lowest values show (critical) factors enhancing and deterring particular or overall sustainability of the evaluated agro-ecosystems.

The results on the integral agricultural sustainability level of this study, based on the micro agro-ecosystem (farm) data, are similar to the previous assessment based on the aggregate sectoral (statistical, etc.) data.

The factors that encourage farming enterprises to improve their economic sustainability include: market demand and price; direct state subsidies; market competition; financial capability; participation in public support programs; the possibility of benefitting immediately; the possibility of benefitting in the near future; tax preferences; the possibility of benefitting in the long term; and closer integration with buyers. The factors that encourage the enhanced social sustainability include: personal convictions and satisfaction; social recognition of individual contributions; immediate benefits for other people and groups; regional community initiatives and pressure; access to advisory services; policies of European Union; and existing regional problems and risks. The factors that encourage farming enterprises to increase their environmental sustainability include: problems and risks existing at the global scale; official regulations, standards and norms; existing regional problems and risks; and European Union policies. All these incentives have to be taken into account in planning the modernization of public policy and programs for sustainable development.

The national and European mechanisms of regulation and support that affect the economic sustainability of the majority of Bulgarian farming enterprises include: direct area-based payments; national top-ups for products and livestock; modernization of agricultural holdings; green payments; and direct support for the semi-subsistence farms. The impact of national and European policies on the social and environmental sustainability of Bulgarian farming enterprises is relatively weak.

There are large differences in the impact of socio-economic, institutional, behavioral, international, natural, etc. factors and individual public policy instruments on the sustainability of farming enterprises of different types and agro-ecosystems.

Having in mind the importance of holistic assessments of this kind for improving the agricultural sustainability, the farm management and the agricultural policies, they should be expended and their precision and representation increased. The latter requires a closer cooperation between and participation of all interested parties, as well as the improvement of the precision through an enlargement of surveyed farms, and incorporating of more "objective" data from field tests and surveys, statistics, expertise of professionals in the area, etc.

REFERENCES

Bachev, H. (2009): *Governing of Agro-ecosystem Services. Modes, Efficiency, Perspectives*, VDM Verlag Dr. Muller Aktiengesellscaft & Co. KG, Saarbrucken.

Bachev H. (2010): *Governance of Agrarian Sustainability*, New York: Nova Science Publishers.

Bachev H (2016): A Framework for Assessing Sustainability of Farming Enterprises. *Journal of Applied Economic Sciences*, Spring Issue, Vol XI, 1(39), 24-43.

Bachev H. (2016): Defining and Assessing the Governance of Agrarian Sustainability, *Journal of Advanced Research in Law and Economics,* Volume VII, Issue 4(18), 797-816.

Bachev H. (2017): Sustainability Level of Bulgarian Farms, *Bulgarian Journal of Agricultural Science*, 23 (1), 1-13.

Bachev H. (2017): Sustainability of Bulgarian Farming Enterprises during EU CAP Implementation, *Journal of Applied Economic Sciences*, 2(48), 422-451.

Bachev H. (2018): *The Sustainability of Farming Enterprises in Bulgaria*, Cambridge Scholars Publishing.

Bachev H. (2018): Institutional Environment and Climate Change Impacts on Sustainability of Bulgarian Agriculture, *Bulgarian Journal of Agricultural Science*, 24 (4), 523-536.

Bachev H., B. Ivanov, D. Toteva, E. Sokolova (2016): Agrarian Sustainability and its Governance – Understanding, Evaluation, Improvement, *Journal of Environmental Management and Tourism*, Vol. 7, issue 4 (16), 639-663.

Bachev H., B. Ivanov, D. Toteva and E. Sokolova (2017): Agrarian sustainability in Bulgaria – economic, social and ecological aspects, *Bulgarian Journal of Agricultural Science*, 23 (4), 519-525.

Bachev H. and D. Terziev (2017): Environmental Sustainability of Agricultural Farms in Bulgaria, *Journal of Environmental Management and Tourism,* Vol 8 No 5 (2017): JEMT Volume VIII Issue 5(21) Fall 2017, 968-994.

Bachev, H., Terziev, D. (2018): A Study on Institutional, Market and Natural Environment Impact on Agrarian Sustainability in Bulgaria, *Journal of Environmental Management and Tourism*, Volume IX, Issue 3 (27), 452-478.

Bachev, H., Terziev, D. (2018). A Study on Agrarian Sustainability Impact of Governance Modes in Bulgaria. *Journal of Applied Economic Sciences,* Volume XIII, Spring, 1(55): 227 - 257.

Belcher K. (1999): *Agroecosystem sustainability: an integrated modelling approach*, PhD Thesis, HARVEST, University of Saskatchewan.

Bohlen P. and G. House (2009): *Sustainable Agroecosystem Management: Integrating Ecology, Economics, and Society*, CRC Press.

De Oliveira A. (editor) (2018): *Sustainability of Agroecosystems*, IntechOpen, DOI: 10.5772/intechopen.70964.

FAO (2013): SAFA. *Sustainability Assessment of Food and Agriculture systems indicators*, FAO.

Fuentes M. (2004): Farms Management Indicators Related to the Policy Dimension in the European Union, *OECD Expert Meeting on Farm Management Indicators and the Environment,* 8-12 March 2004, New Zealand.

Hanna S., I. Osborne-Lee, G. Cesaretti, R. Magdy, T. Khalile (2016): Ecological Agro-ecosystem Sustainable Development in Relationship to Other Sectors in the Economic System, and Human Ecological Footprint and Imprint, *Agriculture and Agricultural Science Procedia*, Volume 8, 17-30.

Hayati D. Z. Ranjbar, and E. Karami (2010): Measuring Agricultural Sustainability, In E. Lichtfouse (ed.), Biodiversity, Biofuels, Agroforestry and Conservation Agriculture, 73, *Sustainable Agriculture Reviews* 5, Springer Science+Business Media B. V., 73-100.

Ikerd J. (2015): *On Defining Sustainable Agriculture*, SARE. http://www.sustainable-ag.ncsu.edu/onsustaibableag.htm.

Ivanov, B., T. Radev, D. Vachevska, P. Borisov (2009): *Agricultural Sustainability* - ASVIWI. Avangard Prima, Sofia.

Lopez-Ridauira S., Masera O., Astier M. (2002): Evaluating the sustainability of complex socio-environmental systems. The MESMIS framework. *Ecological indicators* 2: 135-148.

Millennium Ecosystem Assessment (MEA) (2005): *Ecosystems and Human Well-being: Synthesis*. Washington, DC: Island Press.

Oelbermann M. (editor) (2014): *Sustainable agroecosystems in climate change mitigation,* Wageningen Academic Publishers, doi.org/10.3920/978-90-8686-788-2.

Rezear K., A. Osmani; P. Borisov, D. Skunca (2018): Beyond the Metropolis: Farmers' empowering as a challenge of Peri-urban areas, *European Journal of Economics and Management Sciences*, Vol 3.

Ramírez-Carrillo E., O. López-Corona, J. Toledo-Roy, J. Lovett, F. León-González, L. Osorio-Olvera, J. Equihua, E. Robredo, A. Frank, R. Dirzo, V. Pérez-Cirera (2018): Assessing sustainability in North America's ecosystems using criticality and information theory, *PLOS*, Published: July 16, 2018. https://journals.plos.org/plosone/article?id=10.1371/journal.pone.0200382.

Sauvenier X., J. Valekx, N. Van Cauwenbergh, E. Wauters, H. Bachev. K. Biala, C. Bielders, V. Brouckaert, V. Garcia-Cidad, S. Goyens, M. Hermy, E. Mathijs, B. Muys, M. Vanclooster. and A. Peeters (2005):

Framework for Assessing Sustainability Levels in Belgium Agricultural Systems – SAFE, Belgium Science Policy, Brussels.

Sidle R., W. Benson, J. Carriger, and T. Kamaic (2013): Broader perspective on ecosystem sustainability: Consequences for decision making, *Proc Natl Acad Sci USA*., 110(23): 9201–9208.

Terziev D., D. Radeva, & Y. Kazakova (2018): A new look on agricultural sustainability and food safety: Economic viability, in H. Bachev, S. Che, S. Yancheva (Editors) *Agrarian and Rural Revitalisation Issues in China and Bulgaria*, KSP Books, 231-242.

Todorova K. and R. Treziyska (2018): Agricultural sustainability through provision of agri-environment public goods: The role of farmers as decision-makers, in H. Bachev, S. Che, S. Yancheva (Editors) *Agrarian and Rural Revitalisation Issues in China and Bulgaria*, KSP Books, 253-267.

VanLoon, G., Patil, S., and Hugar, L. (2005): *Agricultural Sustainability: Strategies for Assessment.* London: SAGE Publications.

Zvyatkova D. and A. Sarov (2018): Process of Transfer of Farmily Farms for Sustainability of Agricultural Cooperatives, in *"Role of Family Business for Sustainable Rural Development,"* Agrarian University, 61 (2), 125-134.

In: Agricultural Research Updates. Volume 27 ISBN: 978-1-53615-916-5

Chapter 3

THE EFFECT OF MEASURING THE ECONOMIC AND SOCIAL SUSTAINABILITY INDICES IN PRODUCING CORN USING ANALYSIS NETWORK PROCESS (ANP). THE CASE: VILLAGES OF BUIN ZAHRA IN IRAN

***Somayeh Jangchi Kashani*[1], *Najmeh Daryaei*[1,*] *and Jafar Soltanzad Maleki*[2]**

[1]Department of Agricultural Development,
Science and Research Branch, Islamic Azad University, Tehran, Iran
[2]Department of Architecture, Kerman University, Kerman, Iran

ABSTRACT

Background: Economic and social sustainability dimensions can be used as one of the most important basic factors in policy planning and

[*] Corresponding Author's E-mail: n.daryaei@srbiau.ac.ir.

sustainable production. The complexity and diversity of social and economic indices on the sustainability of production requires considering these indices based on their role and function. This study aimed at identifying and evaluating social and economic sustainability indices in corn production.

Method: Delphi method and Analysis Network Process (ANP) were used to achieve this goal and determine the weight of each index used to establish the economic and social sustainability. The accumulation method had been also based on library studies and interviews with experts and officials.

Results: The research findings indicated that respectively, management dimension (0.467) and participation (0.392) have the maximum weight and as a result, have been the greatest effect between indices of economic and social sustainability.

Conclusion: In other words, the role of management decisions and planning compared to participation and policy-making factors should be highly considered among economic and social sustainability indices in corn production.

Keywords: sustainability, economic, social, impact assessment, corn

1. INTRODUCTION

Sustainability in the production is an important branch of agricultural development. In the 21st century, it is known as the proper coercive in the direction of economic stability and paying attention to the life of the people of the third-world countries. Still agricultural sector is not sustainable enough despite the efforts of third-world countries have been made in this sector and despite the findings of scientific research conducted in this filed. This problem is caused by this major cause that is rooted in how the sustainability concept of the production system in the attitude of these countries. In this regard, studies showed that attitude to the sustainability include mainly technical and environmental dimensions and more attention has been paid to the physical dimension. However, this approach has three dimensions, means economic, social and environmental dimensions according to the relationship between human and nature and then production, and effect of each aspect on a whole produces is known as sustainable system.

Accordingly, both quantitative and qualitative aspects are necessary to meet the objectives of sustainability in the production. Food security plays an important role in the social, economic and political life of country due to climatic conditions and the vulnerability of the agricultural sector in Iran. Corn is one of the crop plants from grain family that its cultivation rising due to the good adaptability with the climatic conditions in most parts of the country. The corn, sugar and starch are among the best crops for grain production due to the high yield grain [1].

The most important reason to choose this product by farmers must be to keep pace with other grain corn production. On the other hand, corn product is competitive compared to other products to increase farmers' income. But in recent years; particularly since the implementation of a plan to increase corn production have been witnessed a surge in the production of this product in the study area. This makes it necessary to examine and design a suitable model to produce corn in a sustainable system more intuitive than ever. Considering the importance of corn production in the study area, this research focuses on the design of a sustainable system of production in the Qazvin. As well as this research will examine the requirements and challenges encountering the corn production in the form of a sustainable production system.

1.1. Theoretical Basics

Analytic Hierarchy Process (AHP) is one of the most effective methods to make a decision that was first introduced by L. Saaty (1999) [2]. This method is based on paired comparisons and enables managers to evaluate different scenarios. The ability to analyze a decision-making problem with a tiered structure is basic infrastructure in the use of AHP method. A method known as Analytic Network Process (ANP) was provided based on super matrix approach due to the broad spectrum of decision feedback systems and to reform AHP. This method can be used to complex issues of non-class structure. The possibility of making the reciprocal relationship between different levels of decision- making and internal communication of decision indices together in same level are the major advantages of this method. The unique feature of ANP method compared to the AHP, cause a higher tendency of most researchers and managers to it [3]. Van Passel et al. (2007) identified that large farms are sustained at a higher level in a study entitled farm sustainability assessment using ANP. Age of farmers and dependence on support payment creates the difference in performance in achieving sustainability [4]. Farajzadeh & Takloobighash (2001) prepared the Agro-climatic zoning of Hamedan by ANP technique and with an emphasis on wheat. The physical factors and climatic elements such as topography, slope, soil type, soil depth, soil texture, annual precipitation, annual temperature and evaporation analyzed and measured in this study. Then the final map of suitable areas to corn cultivation was provided by above layers overlap [5]. Ghaneikhoo (2001) examined agricultural climate of Firozabad by emphasizing on corn using a hierarchical approach in his master's thesis. The effect of climatic factors such as temperature, cloudiness, humidity, wind, radiation and the other environmental factors such as water quality, soil quality and poor quality of water and soil in corn cultivation was assessed and the best times and places for corn cultivation were determined in this research [6]. Adeli et al. (2011) examined the factors affecting the sustainability of agricultural operations using network analysis. They concluded that the mean of sustainability of the agricultural practices in 52.3 percent of cases was less and was higher than total mean of sustainability index in 47.7 percent [7].

There is no operation and management method formally about the economic and social analysis of corn cultivated in Iran according to the research that has been conducted so far. Thus, this study is planned to analyze the model in two economic and social indices for the corn production with sustainability approach. For this purpose, the Buin Zahra will be discussed as a case study to design the model.

2. Materials and Methods

It will be discussed to analyze the factors affecting assessment of the economic and social consequences to produce the corn in Buin Zahra villages by Analysis Network Process (ANP) and applying this technique as a sub-category of the Multi-index Decision-Making (MCDM) in this study. For this purpose, the necessary information are collected using a questionnaire, and pattern is then analyzed. In this research, Delphi technique is used to identify and screen the most important decision-making indices. Delphi is a method based on the intuitive perspective of the experts, in which a group of experts reaches a consensus after expressing their perspectives on a specific issue. Although this judgment seems to be a mental, it is more credible than individual statements, because it provides more objective results [8]. In the Delphi method, the questioning is performed in two steps or more, and the results resulted from the previous step or steps are used at each stage. Therefore, experts respond to questions from the perspectives and opinions of their peers and the results from the previous stage in the second stage. It should be noted that the apparent validity of the questionnaires of each stage was verified by some university professors and experts in the field of agriculture. Dalki (1969) expresses that the reliability of questionnaires in this technique indicates that the coefficient of reliability will be more than 0.80 (≥ 0.8) if the size of the Delphi group is more than 13[9]. The statistical population of this study was 30 experts in the oil seeds department of the Agricultural organization in Buin Zahra.

ANP is one of the most effective techniques to make a decision with multiple indices that were first introduced in 1982 by Thomas L. Saaty and as a developed form of AHP method. ANP improves AHP as an instrument of multi-index decision-making by replacement network instead of hierarchy. ANP process has four main steps [10]. The first step is to create model and formulate problem. In the next stage of modeling, logical system of a network must be clearly expressed and analyzed. . An objective of decision-making, decision-making indices and possible options may be specified at this stage. This structure can be created with the opinions of decision-makers. In AHP, there is a linear and unidirectional relationship between the purpose, index, sub-index and options. Option does not affect the index, and the indices do not affect each other [11]. The open network structure of ANP provides the possibility modeling complex relationships between different levels of decision and indices. Network model is composed of a combination of 4 levels (Figure 1) that is known sub-network ANP in this study. Super Decisions software v. 2.6.0-RC1 has been used to analyze the information in this study.

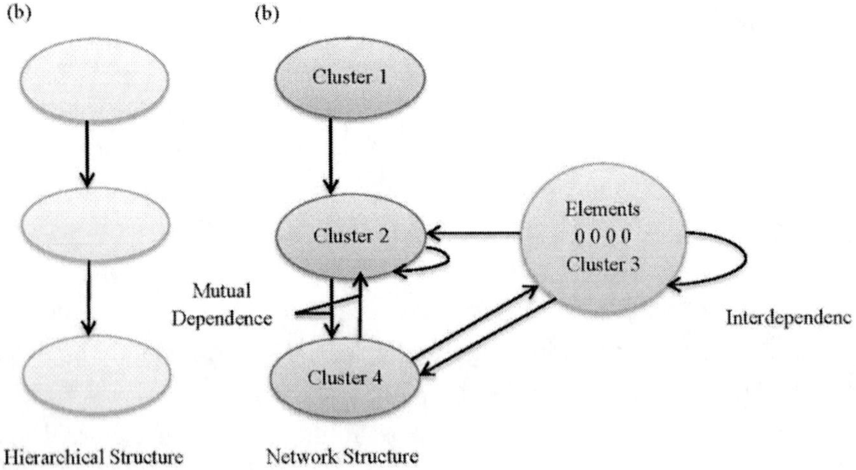

Figure 1. Diagram of the structural difference between a hierarchy and network [12].

2.1. Research Model

The first step of matrix: the indices on the first level, categories on the second level and sub-index (variables) are located in the third level (Figure 2). The first one was referred to specialists (university professors and experts) to design the model. A questionnaire about explore factors influencing economic and social consequences of corn was developed and delivered to the sample. The index and sub-index are distinguished using survey results.

The second step of matrix: binary compare and calculate the vectors of priority. At this stage, a set of paired comparisons was calculated to estimate the relative importance of each of the factors and indices that are effective in target selection. If there is interconnectedness of the elements of a component, binary comparisons should be used.

The relative importance is determined by the scale 1 to 9 where a score 1 indicates the importance of the two factors, while a score 9 indicates the absolute importance of a factor (element row) compared to other factors (element column) (Table 1). As well as mutual sizes can be considered to reversed compare, such that aij = 1/aij (aij represents the importance of i element/j element). This kind of comparison is according to the proposed method of Saaty (Saaty, 1996).

The pairwise comparison in ANP such as AHP is made within a matrix. By the following equation, a priority vector w has been calculated as an estimate of the relative importance of compared factors or elements:

$$AW = \lambda_{\max} w$$

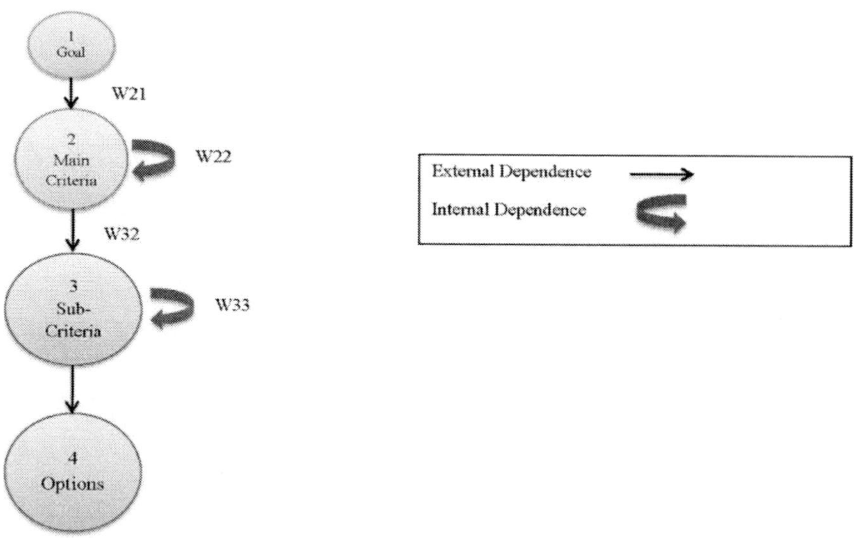

Figure 2. The network model designed to assess the economic and social consequences of corn cultivation.

Table 1. Size of preference for binary comparison

Preferences (judgment about the importance)	Numerical Value
Same Preference	1
Same to Low Preference	2
Low Preference (bit absolute)	3
Low to High Preference	4
High Preference	5
High to Very High Preference	6
Very High Preference	7
Very High to Absolute Preference	8
Absolute Preference	9

A, pairwise comparison matrix and w, weight vector and λmax is the biggest weight vector of matrix A. Priority vectors should be calculated for all pairwise comparison matrices. Catch rate (C.R) should be calculated for each matrix to determine the compatibility comparisons. If C.R < 0.1; comparisons to be accepted as a compatible comparison [13], and if not; otherwise it is better for decision-makers to reconsider their judgments. The rate adjustment can be calculated for a comparison matrix by the following relationship:

$$\mathrm{CR} = \frac{CI}{RI}$$

Where consistency index (CI) shows a measure of deviation from the adjustment and obtained from following equation:

$$\mathrm{CI} = \frac{\lambda_{max} \ -n}{n-1}$$

λmax is the maximum eigenvalues of comparisons matrix and n is the length of binary comparison matrix. (RI) is the random consistency index or the average weight that is randomly provided. The value of this index is based on the dimensions of comparison (n*n). Group decision-making is used to avoid any biased attitude in binary comparisons matrix. There are several ways to bring the attitudes and prejudices of the members in the binary comparison matrix. Saaty showed that the geometric mean of individual judgment is the best approach to use. The geometric mean is mathematically the best mean, due to maintain the opposite effect on binary comparison matrix. If we assume that $a_{ij}^{(k)}$ is item related to the K component to compare i factor to j, then the geometric mean is calculated as following [10]:

$$a_{ij} = \sqrt[N]{\prod_{k=1}^{N} a_{ij}^{(k)}}$$

The above equation is set when the attitudes of group members have the same importance degree.

The third step is to form Super Matrix: the concept of the super matrix is similar to the Markov chain process [14]. The topical priority vectors can be used to fit in the columns of the matrix to achieve the final military superiority that affected the inner dependency. This matrix is known as the "Super matrix". Therefore, a super matrix is practically a segmentation matrix which each part of them indicates the relationship among two nodes (component or category) in a system [15].

If the components of a decision-making system is C_N (N = 1,...n), each component of N will have m_N that is shown as $e_{N1}, e_{N2}, \dots, e_{NnN}$. The priority vectors obtained in second step have been categorized based on the effect that a component has on itself or another ones, the matrix is placed in the right spots. The standard form of a super matrix is shown in the following (Figure 3).

In this study, the super matrix for a three-level structure of the network is as follows.

$$W_n = \begin{bmatrix} 0 & 0 & 0 \\ w_{21} & w_{22} & 0 \\ 0 & w_{32} & I \end{bmatrix}$$

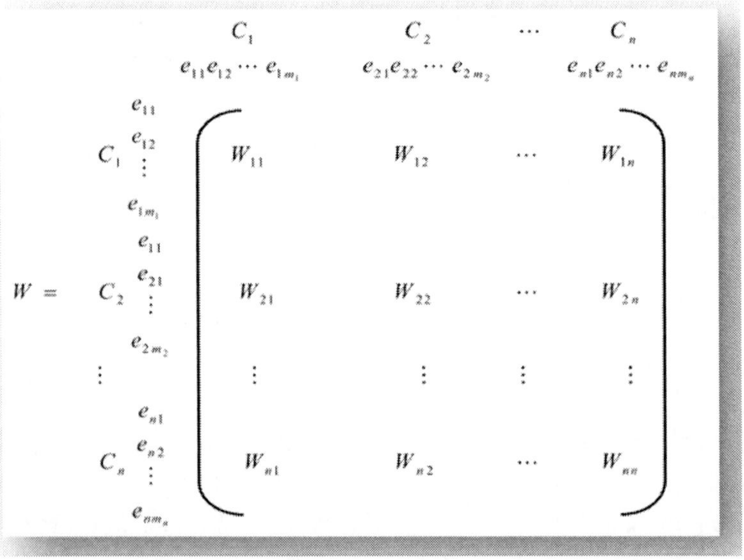

Figure 3. The standard form of a super matrix.

The vector of W_{21} reflects the effect of target on benchmark. The matrix W_{32} represents the effect of index on the sub-index and I is the unite matrix. Zero will show ineffective corresponding element that can be replaced in a component or among two components, if there is internal communication. Interdependencies of index also shown by W_{22} .

In the process model of network analysis, the super matrix is calculated in three stages. Calculated weight for all binary comparisons will be first entered in the matrix that so-called unweight super matrix. In the next stage, the weighted super matrix will be calculated by multiplying the size of the non-weighted super matrix in corresponding sizes of comparison matrix in index. Then the weighted matrix sizes are standard, so that the total size of the columns of the matrix is equal to 1. Super matrix is calculated within a matrix in the third and final stage where the columns are the same size [10]. Saaty confirmed that the final weight is obtained from the following equation [16] using probable matrix and Markov chains:

$$\lim_{k \to \infty} W^{2k+1} = W_c$$

Where W is the limit matrix (final weight) and W_C is the standardized weighted matrix. The fourth step is to choose the best options. If the super matrix is formed in the third step to cover the entire network, lead weights in column options can be found in super normalized matrix. On the other hand; if a super matrix contains only components that are interrelated, additional calculations must be done to obtain final priority of options. The highest priority with quantity option should be the first choice.

3. Results

3.1. Implementing the Process Model in Network Analysis in this Case

Stage 1: We need to be formed to a suitable network model for research purpose in order to run a network analysis model (measuring the economic and social sustainability indices), followed by principal components, index and sub-index (Table 2).

Figure 4 shows a network model to evaluate the economic and social sustainability indices.

The priority vectors are calculated at this stage. Also the binary comparing at this stage has been done to obtain the relative importance of each factor and indices that are effective in target selection. If there is interconnectedness of the elements of a component, binary comparisons should be used. The sizes of relative importance will be determined by the scale of 1 to 9 where a score of 1 indicates prospective importance of the two factors, while a score of 9 indicates the absolute importance of a factor (element of row) compared to other factors (element of column).

Stage 2: we move to the indices after forming the dependency tree among variables that begins the target. The overall structure of the matrix or the unbalanced super matrixes are also identified in this step with regard to the structure of the network model. There is noticeable in Figure 5 that the C indicates the nodes and e represents the elements in the nodes.

Vectors W within the matrix is the weight vectors of paired comparisons with each other node elements [17].

Table 2. index, sub-index and selected options

Indices	Indices Code	Sub-index	Sub-index Code
Economic Factors	Ef	Reduce intermediaries by buying corn	Ec1
		Guaranteed purchase of corn by cooperatives	Ec2
		Stabilize prices and prevent fluctuations in different years	Ec3
		Financial support of corn farmers by government (subsidies, loans, etc.)	Ec4
		Insurance products	Ec5
		Insurance marketing stages (transportation, warehousing, etc.)	Ec6
		Set a reasonable price for product	Ec7
		Corn farmers access to natural inputs	Ec8
		Access to international markets for export corn	Ec9
		Marketing possibility to sell abroad	Ec10
		Define and supply appropriate subsidies to provide inputs and necessary tools to stabilize corn production	Ec11
		According to the development of healthy processing industries from corn in order to increase the added value	Ec12
Social Factors	Sf	The use of indigenous knowledge of local farmers in sustainable cultivation of corn	So1
		Create a positive vision and belief in corn farmers about the sustainable production	So2
		The participation of the private sector and using their capabilities	So3
		Create cooperatives in the adoption and implementation of sustainable production	So4
		Create local institutions according to the needs of corn farmers	So5
		Collaborating corn farmers with extension agents in applying principles of corn sustainable production	So6
		Collaborating corn farmers with researchers of research institutes in implementing research projects in sustainable production of corn	So7
		Appropriate conditions for corn farmers participation in the decision-making and planning for sustainable production of corn	So8
		Culture-building and advertising for the use of corn by consumers	So9

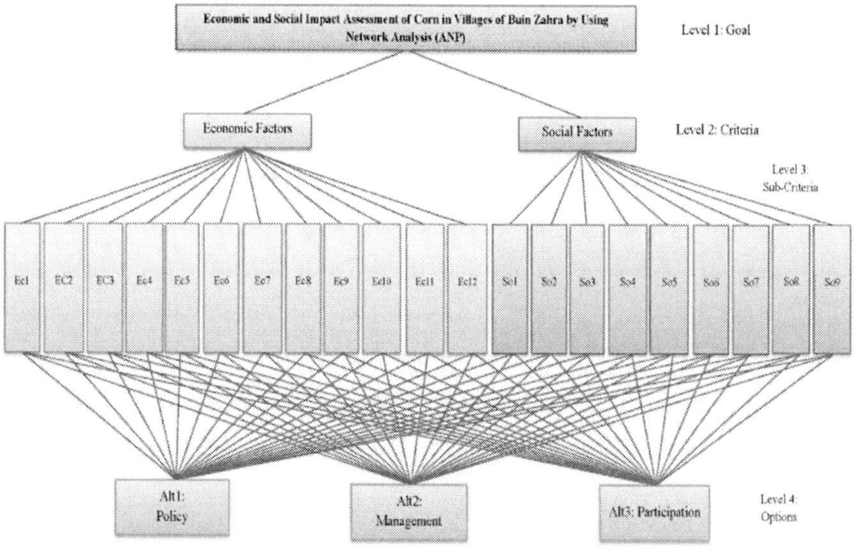

Figure 4. Schematic network model to assess the economic and social sustainability indices.

		C_1				C_2						C_N			
		e_{11}	e_{12}	...	e_{1N}	e_{21}	e_{22}	...	e_{2N}	...		e_{N1}	e_{N2}	...	e_{NN}
C_1	e_{11}	W_{11}				W_{12}				...		W_{1N}			
	e_{12}														
	...														
	e_{1N}														
C_2	e_{21}	W_{21}				W_{22}				...		W_{2N}			
	e_{22}														
	...														
	e_{2N}														
⋮		...				...				...		...			
C_N	e_{N1}	W_{N1}				W_{N2}				...		W_{NN}			

Figure 5. The overall structure of super matrixes [14].

As comparative matrix of alternatives was evaluated, alternatives dependence on each other, index, and indices dependence on each other, sub-index and sub-index dependence on each other, as well as their compatibility are also controlled. Then, a binary comparison of three options has been done based on the scale 9 of Saati's quantity in the same way that the Analytic Hierarchy Process (AHP) is used. The result of binary comparison of alternatives and weighted vector W_{21} is presented in Table 3. Group judgment is used to binary comparison of principal index to achieve a desired result that the elements in binary comparison matrix of indices are obtained by the mean of the group viewpoints.

Stage 3: At first, we characterized the interaction among the components. It is also possible that there is no relationship between the components. However if there is any connection, this relationship can be one-way or two-way. As you can see in Table 4, the relationship between the components in this study is two-way.

Now binary comparison between them based on the scale 9 of Saati's quantity has done in order to achieve elements of matrix W_{22} and to understand the interdependencies between components. The binary comparison of other two components of management and participation (by controlling the components of the policy) is provided in the Table 5 to calculate the importance coefficient of each component (due to the interdependence between them).

Similarly, interdependence and mutual of components are examined by controlling the other two components. As well as three matrix of binary comparison are formed when each one is controlled by the adjustment factor. This factor must always be less than 0.01 to be calculated the matrix of interdependencies for the main index (W_{22}). The results are presented in Table 6 after the formation of the three matrices and necessary calculations.

Table 3. Binary comparison and special values of ternary components

Items	Management	Participation	Policy	Special vector S, E
Management	1	4.5	4	0.467
Participation	3	1	2	0.392
Policy	2	1.5	1	0.139

Table 4. alternatives interdependence with each other

Items	Management	Participation	Policy
Management			
Participation			
Policy			

Table 5. Binary comparison of items due to their interdependence with control policy item

Items	Management	Participation	Special vector W
Management	1	21/10	0.643
Participation	10/21	1	0.221

Table 6. Matrix W_{22}

Items	Management	Participation	Policy
Management	0	0.270	0.643
Participation	0.241	0	0.221
Policy	0.174	0.102	0

Table 7. Interdependence of the economic sub-index with each other

Economic Variables	Ec1	Ec2	Ec3	Ec4	Ec5	Ec6	Ec7	Ec8	Ec9	Ec10	Ec11	Ec12
Ec1		*	*	*			*		*			
Ec2	*		*	*	*		*	*	*		*	
Ec3	*	*		*			*				*	
Ec4	*	*	*		*	*	*	*	*	*	*	*
Ec5		*	*	*		*			*	*		
Ec6		*	*	*	*				*	*		
Ec7	*	*	*	*	*	*		*	*	*	*	*
Ec8		*	*	*	*	*						
Ec9	*	*	*	*	*	*	*			*		*
Ec10	*	*	*	*		*	*					*
Ec11			*	*	*		*	*				*
Ec12	*			*			*		*	*		

Table 8. Interdependence matrix of the social index

Social Variables	So1	So2	So3	So4	So5	So6	So7	So8	So9
So1		*	*		*	*			
So2			*	*	*	*	*		*
So3	*	*			*	*	*	*	*
So4	*	*	*		*	*	*	*	*
So5	*	*	*	*		*	*	*	
So6	*	*	*		*		*	*	*
So7	*	*		*	*	*		*	
So8			*	*	*	*	*		
So9		*	*	*	*	*	*		

Step 4: 21 index that reflect the economic (12 index) and social (9 index) characteristics of the component were selected as the network model has been revealed, for the purposes of this study. Then three alternatives including management, participation and policy are intended to assess each criterion. Expert's viewpoints which are shown in Tables 7 & 8 are used to provide this table and to determine sub-index interdependencies.

As can be seen, all the sub-index have mutual effects on each other.

Two other matrices (W_{43} and W_{44}) can be calculated according to primary super matrix at this stage. Uneven super matrix can be created that can be seen in the Tables 9 and 10 by replacing these matrices in the primary super matrix, created. Uneven super matrices must be converted to the harmonious super matrices (what it is called random matrix by Saaty). The sum of the values of each column of the matrix is equal to 1. Uneven super matrix must be multiplied in a cluster matrix in order to become an uneven matrix to the harmonious super matrix (Tables 11 & 12).

The objective of limiting the harmonious super matrix is to achieve the long-term relative effect of each element on each other (Tables 13, 14, 15 & 16). Each element of the harmonious matrix can struggle to be K (a number is optional) to divergence of the impact coefficient [12].

Table 9. Uneven super matrix of the economic sub-index

	Ec1	Ec2	Ec3	Ec4	Ec5	Ec6	Ec7	Ec8	Ec9	Ec10	Ec11	Ec12
Ec1	0.1728	0.1006	0.1244	0.1618	0.0678	0.1062	0.1009	0.1545	0.1383	0.0930	0.0936	0.1039
Ec2	0.1128	0.1902	0.1752	0.1217	0.1426	0.1514	0.1295	0.1000	0.0947	0.1057	0.1036	0.0769
Ec3	0.0598	0.0602	0.0759	0.0766	0.0587	0.0631	0.0838	0.0605	0.0863	0.1167	0.0848	0.0751
Ec4	0.0853	0.1003	0.1002	0.0682	0.1463	0.0522	0.0840	0.0777	0.0841	0.0803	0.0682	0.0868
Ec5	0.0819	0.0911	0.0533	0.0636	0.0472	0.0219	0.0696	0.0257	0.0861	0.0795	0.0601	0.0692
Ec6	0.0481	0.0476	0.0800	0.0505	0.0185	0.0845	0.0468	0.1336	0.0631	0.0642	0.0600	0.0582
Ec7	0.0729	0.0454	0.0477	0.0240	0.0751	0.0498	0.0617	0.0495	0.0524	0.0528	0.0950	0.0720
Ec8	0.0416	0.0421	0.0356	0.0202	0.0356	0.0310	0.0615	0.0589	0.0471	0.0395	0.0557	0.0700
Ec9	0.0346	0.0299	0.0292	0.0180	0.0374	0.0339	0.0534	0.0419	0.0428	0.0431	0.0417	0.0457
Ec10	0.0218	0.0192	0.0236	0.0138	0.0295	0.0412	0.0309	0.0369	0.0355	0.0374	0.0424	0.0490
Ec11	0.0300	0.0182	0.0144	0.0361	0.1028	0.0218	0.0303	0.0163	0.0257	0.0316	0.0318	0.0375
Ec12	0.0160	0.0328	0.0183	0.1234	0.0164	0.1208	0.0255	0.0223	0.0217	0.0339	0.0410	0.0335

Table 10. Uneven super matrix of the social sub-index

	So1	So2	So3	So4	So5	So6	So7	So8	So9
So1	0.1522	0.1360	0.1505	0.1613	0.1671	0.0973	0.1487	0.1719	0.1492
So2	0.1287	0.0881	0.1062	0.1376	0.1471	0.1546	0.1135	0.1369	0.1335
So3	0.1092	0.1144	0.1075	0.1070	0.1052	0.1144	0.0729	0.1029	0.1260
So4	0.0777	0.1116	0.0974	0.0934	0.0724	0.1005	0.0832	0.0664	0.0888
So5	0.0716	0.0735	0.0723	0.0681	0.0687	0.0747	0.0788	0.0665	0.0654
So6	0.0580	0.0640	0.0648	0.0579	0.0504	0.0660	0.0642	0.0663	0.0563
So7	0.0577	0.0611	0.0535	0.0457	0.0492	0.0526	0.0953	0.0428	0.0520
So8	0.0414	0.0400	0.0528	0.0329	0.0352	0.0395	0.0478	0.0472	0.0325
So9	0.0405	0.0481	0.0320	0.0330	0.0415	0.0373	0.0325	0.0358	0.0331

Table 11. Harmonious super matrix of the economic sub-index

		Ec1	Ec2	Ec3	Ec4	Ec5	Ec6	Ec7	Ec8	Ec9	Ec10	Ec11	Ec12
subcritria1	Ec1	0.2221	0.1293	0.1600	0.2081	0.0871	0.1366	0.1297	0.1986	0.1778	0.1196	0.1203	0.1336
	Ec2	0.1451	0.2445	0.2252	0.1564	0.1833	0.1946	0.1665	0.1286	0.1218	0.1359	0.1332	0.0989
	Ec3	0.0768	0.0774	0.0976	0.0985	0.0755	0.0811	0.1077	0.0778	0.1110	0.1501	0.1090	0.0966
	Ec4	0.1097	0.1290	0.1289	0.0876	0.1881	0.0671	0.1080	0.0999	0.1081	0.1032	0.0877	0.1116
	Ec5	0.1053	0.1171	0.0685	0.0818	0.0606	0.0281	0.0894	0.0331	0.1107	0.1022	0.0773	0.0890
	Ec6	0.0619	0.0612	0.1029	0.0650	0.0238	0.1086	0.0602	0.1718	0.0811	0.0826	0.0771	0.0748
	Ec7	0.0938	0.0584	0.0613	0.0308	0.0965	0.0640	0.0794	0.0637	0.0673	0.0679	0.1221	0.0926
	Ec8	0.0535	0.0542	0.0458	0.0259	0.0458	0.0399	0.0791	0.0757	0.0605	0.0508	0.0716	0.0900
	Ec9	0.0445	0.0385	0.0375	0.0232	0.0481	0.0436	0.0686	0.0539	0.0551	0.0554	0.0536	0.0587
	Ec10	0.0281	0.0247	0.0303	0.0177	0.0380	0.0529	0.0397	0.0474	0.0456	0.0481	0.0545	0.0630
	Ec11	0.0386	0.0235	0.0185	0.0464	0.1322	0.0281	0.0390	0.0209	0.0331	0.0406	0.0409	0.0483
	Ec12	0.0206	0.0422	0.0236	0.1586	0.0210	0.1554	0.0328	0.0287	0.0279	0.0436	0.0527	0.0431

Table 12. Harmonious super matrix of the social sub-index

		So1	So2	So3	So4	So5	So6	So7	So8	So9
subcritria2	So1	0.2065	0.1846	0.2042	0.2189	0.2268	0.1321	0.2018	0.2333	0.2025
	So2	0.1746	0.1196	0.1441	0.1867	0.1996	0.2098	0.1540	0.1857	0.1812
	So3	0.1482	0.1553	0.1458	0.1453	0.1428	0.1552	0.0990	0.1397	0.1711
	So4	0.1054	0.1515	0.1321	0.1268	0.0982	0.1364	0.1129	0.0901	0.1206
	So5	0.0972	0.0998	0.0981	0.0924	0.0932	0.1014	0.1070	0.0903	0.0887
	So6	0.0787	0.0868	0.0880	0.0785	0.0684	0.0896	0.0871	0.0900	0.0764
	So7	0.0783	0.0829	0.0726	0.0620	0.0668	0.0713	0.1293	0.0581	0.0705
	So8	0.0562	0.0543	0.0717	0.0447	0.0478	0.0536	0.0648	0.0641	0.0441
	So9	0.0549	0.0653	0.0434	0.0448	0.0563	0.0506	0.0442	0.0486	0.0449

Table 13. Super matrix of sub-index limit (economic factors)

	Ec1	Ec2	Ec3	Ec4	Ec5	Ec6	Ec7	Ec8	Ec9	Ec10	Ec11	Ec12
Ec1	0.123029	0.123029	0.123029	0.123029	0.123029	0.123029	0.123029	0.123029	0.123029	0.123029	0.123029	0.123029
Ec2	0.135401	0.135401	0.135401	0.135401	0.135401	0.135401	0.135401	0.135401	0.135401	0.135401	0.135401	0.135401
Ec3	0.070265	0.070265	0.070265	0.070265	0.070265	0.070265	0.070265	0.070265	0.070265	0.070265	0.070265	0.070265
Ec4	0.088638	0.088638	0.088638	0.088638	0.088638	0.088638	0.088638	0.088638	0.088638	0.088638	0.088638	0.088638
Ec5	0.065961	0.065961	0.065961	0.065961	0.065961	0.065961	0.065961	0.065961	0.065961	0.065961	0.065961	0.065961
Ec6	0.058231	0.058231	0.058231	0.058231	0.058231	0.058231	0.058231	0.058231	0.058231	0.058231	0.058231	0.058231
Ec7	0.055886	0.055886	0.055886	0.055886	0.055886	0.055886	0.055886	0.055886	0.055886	0.055886	0.055886	0.055886
Ec8	0.042114	0.042114	0.042114	0.042114	0.042114	0.042114	0.042114	0.042114	0.042114	0.042114	0.042114	0.042114
Ec9	0.034915	0.034915	0.034915	0.034915	0.034915	0.034915	0.034915	0.034915	0.034915	0.034915	0.034915	0.034915
Ec10	0.027718	0.027718	0.027718	0.027718	0.027718	0.027718	0.027718	0.027718	0.027718	0.027718	0.027718	0.027718
Ec11	0.032443	0.032443	0.032443	0.032443	0.032443	0.032443	0.032443	0.032443	0.032443	0.032443	0.032443	0.032443
Ec12	0.043178	0.043178	0.043178	0.043178	0.043178	0.043178	0.043178	0.043178	0.043178	0.043178	0.043178	0.043178

Table 14. Super matrix of sub-index limit (social factors)

	So1	So2	So3	So4	So5	So6	So7	So8	So9
So1	0.14795	0.14795	0.14795	0.14795	0.14795	0.14795	0.14795	0.14795	0.14795
So2	0.123159	0.123159	0.123159	0.123159	0.123159	0.123159	0.123159	0.123159	0.123159
So3	0.107321	0.107321	0.107321	0.107321	0.107321	0.107321	0.107321	0.107321	0.107321
So4	0.089862	0.089862	0.089862	0.089862	0.089862	0.089862	0.089862	0.089862	0.089862
So5	0.071519	0.071519	0.071519	0.071519	0.071519	0.071519	0.071519	0.071519	0.071519
So6	0.060746	0.060746	0.060746	0.060746	0.060746	0.060746	0.060746	0.060746	0.060746
So7	0.056689	0.056689	0.056689	0.056689	0.056689	0.056689	0.056689	0.056689	0.056689
So8	0.041403	0.041403	0.041403	0.041403	0.041403	0.041403	0.041403	0.041403	0.041403
So9	0.038194	0.038194	0.038194	0.038194	0.038194	0.038194	0.038194	0.038194	0.038194

Table 15. Super matrix of sub-index limit based on the sub-index importance and effect (economic factors) for the alternatives

	Ec1	Ec2	Ec3	Ec4	Ec5	Ec6	Ec7	Ec8	Ec9	Ec10	Ec11	Ec12
Management	0.064145	0.064145	0.064145	0.064145	0.064145	0.064145	0.064145	0.064145	0.064145	0.064145	0.064145	0.064145
Participation	0.021579	0.021579	0.021579	0.021579	0.021579	0.021579	0.021579	0.021579	0.021579	0.021579	0.021579	0.021579
Policy	0.136498	0.136498	0.136498	0.136498	0.136498	0.136498	0.136498	0.136498	0.136498	0.136498	0.136498	0.136498

Table 16. Super matrix of sub-index limit based on the sub-index importance and effect (social factors) for the alternatives

	So1	So2	So3	So4	So5	So6	So7	So8	So9
Management	0.083185	0.083185	0.083185	0.083185	0.083185	0.083185	0.083185	0.083185	0.083185
Participation	0.132062	0.132062	0.132062	0.132062	0.132062	0.132062	0.132062	0.132062	0.132062
Policy	0.047911	0.047911	0.047911	0.047911	0.047911	0.047911	0.047911	0.047911	0.047911

The model of network analysis has identified the weight of each index. According to this index; management (0.467), participation (0.392) and policy (0.139) factors have the highest importance and thus highest effect on meeting the research objective (Figure 5).

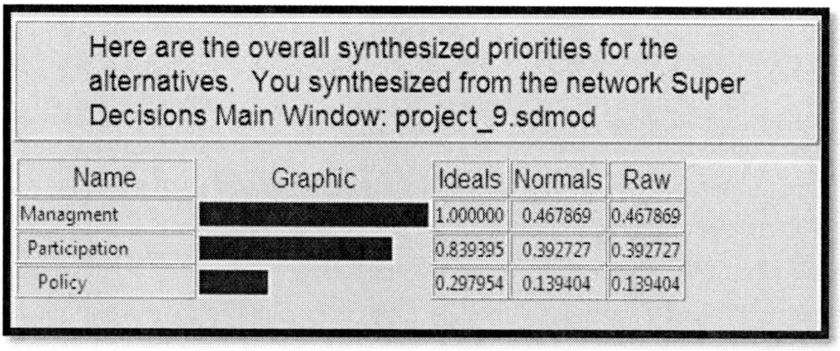

Figure 5. Overall synthesized priorities for the alternatives.

4. Discussion and Conclusion

If we accept the sustainable development as a way to meet the present needs without compromising the ability of future generations (reported by Brandt Land), then it cannot be imagined without social and economic variables. For example, if there is no attention to policy-making, management and participation issues, how can we be hopeful for sustain production and use of resources for future generations. It can be said that sustainable development caused by social and economic sustainability and the underlying achieve sustainable development pays attention to economic and social issues in addition to environmental issues. The theoretical principles are used to meet social and economic sustainability in this survey, the three components of management, participation and policy-making are introduced. These three components are in the form of 21 indices (12 economic indices) and (9 social Index). On the other hand, the desired indices and components in this study rated the inner and interdependence with each other. Since the issues that somehow show the reciprocal relationship among the indexes and there are internal, the use of hierarchical analysis considers the relationship among them and the result is wrong. The network analysis has been used in this case. The ANP will not affect output, even if the indices are interrelated. Therefore ANP has been used in a proper finishing result according to the mutual relationship between the economic and social sustainability indices.

According to the research findings, management has been allocated most weight to itself (0.467) among other social and economic cases. On the other hand, if we accept economic and social sustainability as an integral component to meet sustainable development of corn product, it can be said that the management is most important factor to reach the social and economic sustainability in producing this product. In fact, sustainable development requires proper management and planning in economic and social dimensions. The indices that have been used in this research are definitions and perspectives of researchers and pundits about the subject of the social and economic sustainability in producing corn. Therefore, the economic and social sustainability indices can be measured for the other products with new items for each index in the future research. The weight of these indices showed the realization of ultimate objective or the economic and social sustainability of corn on a priority basis. Management and participation can have high importance factors in achieving the economic and social sustainability of corn production according to the findings of this research.

REFERENCES

[1] Akhavan, K; Shiri, M; Kazemi Azar, F. The effect of drip irrigation water and planting arrangement on corn yield, *Journal of Water Research in Agriculture*, **28**(1), 97-105, (2014), (In Persian).

[2] Saaty, TL. *Fundamentals of the analytic network process*, ISAHP 1999, Kobe, Japan, August 12-14, pp. 1-14 (1999).

[3] Samani, JM; Delavare, M. Application of Network Analysis Process (ANP) in prioritizing shrimp breeding plants, *Journal of Iranian Water Resources Research*, **6**(2), 46-54, (2010), (In Persian).

[4] Van Passel, S; Nevens, F; Mathijs, E; Van Huylenbroeck, G. Measuring farm sustainability and explaining differences in sustainable efficiency, *Journal of Ecological Economics*, **62**(1), 149-161, (2007).

[5] Farajzadeh, M; Takloobighash, A. Agro-climate distribution of Hamedan province using geographical information system with emphasis on dried Wheat, *Quarterly Journal of Geographical Research*, **41**(33), 93-105, (2001), (In Persian).

[6] Ghaneikhoo, M. Agricultural climate of Firoozabad city with an emphasis on Corn cultivation, Master's Thesis in *Natural Geography*, Isfahan University, (2001), (In Persian).

[7] Adeli Sardooei, M; Hayati, B; Zarifian, S; Hosseini-Nasab, D. Factors affecting agricultural operation sustainability in Jiroft city (case study: Onion, Potatoes and Tomatoes), *Journal of Agriculture Economics and Development (Agriculture Sciences and Technology)*, **25**(4), 459-468, (2011), (In Persian).

[8] Pashaeizad, H. A glance to Delphi method, *Journal of Peyke-e-Noor*, **6**(2), 63-79, (2007), (In Persian).

[9] Pezeshkirad, GH; Feli, S. Challenges and solutions for processing, marketing and exporting Saffron: using Delphi techniques, *Journal of Agricultural Economics*, **4**(1), 137-157, (2010), (In Persian).

[10] Saaty, TL. *Theory and Application of the Analytic Network Process: Decision Making with Benefits, opportunities, Costs and Risks* (3 edition), RWS Publications, New York, 353 pages, (2005).

[11] Kurttila, M; Pesonn, M; Kangas, J; Kajanus, M. Utilizing the analytic hierarchy process in SWOT analysis: A hybrid method and its application to a forest – certification case, *Journal of Forest Policy and Economics*, **1**, 41-52, (2000).

[12] Zebardast, E. The application of Analytic Network Process (ANP) in urban and regional planning, *Journal of Honar-ha-ye-ziba (Memari-va-Shahrsazi)*, **41**, 79-90, (2010), (In Persian).

[13] Azar, A; Rajabzadeh, A. *Applied Decision Making (MADM Approach)*, Negah-e-Danesh, Tehran, 184 pages, (2002), (In Persian).

[14] Saaty, TL. *Decision Making with Dependence and Feedback: The analytical network process: the organization and prioritization of complexity*, RWS publications, New York, 370 pages, (1996).

[15] Meade, LM; Sarkis, j. Analyzing organizational project alternatives for agile manufacturing processes: An analytical network approach, *International Journal of Production Research*, **37**(2), 241-261, (1999).

[16] Ghodsipour, H. *Analytical Hierarchy Process*, Amir Kabir University, Tehran, 230 pages (2008), (In Persian).

[17] Dorri, B; Hamzei, E. Determining risk response strategy in risk management by ANP technique (case study: North Azadgan field development project), *Journal of Industrial Management*, **2**(4), 75-92, (2010), (In Persian).

In: Agricultural Research Updates. Volume 27 ISBN: 978-1-53615-916-5
Editors: Prathamesh Gorawala et al.

Chapter 4

COMPARISON BETWEEN STATISTICAL PRECIPITATION PREDICTION IN NORTHERN PATAGONIA (ARGENTINA) USING ERA - INTERIM AND NCEP REANALYSIS DATASETS

Marcela Hebe González[1,2,*] and Alfredo Luis Rolla[2]

[1]Departamento de Ciencias de la Atmósfera y los Océanos,
Facultad de Ciencias Exactas y Naturales,
Universidad de Buenos Aires, Buenos Aires, Argentina
[2]Centro de Investigaciones del Mar y la Atmósfera
(CONICET-UBA)
Buenos Aires, Argentina

---

[*]Corresponding Author's E-mail: gonzalez@cima.fcen.uba.ar.

Abstract

Northern Patagonia is a dry area with scarce precipitation and agricultural and livestock activities which are highly dependent on inter-annual rainfall variability. In this study statistical regression models were derived to estimate summer precipitation in a single location using two different sets of data: ERA - INTERIM and NCEP reanalysis. Both datasets were used to determine the predictors. The models were derived using lineal backward regression techniques and the different efficiency between the derived models was analyzed. Results pointed out the best resolution of ERA-INTERIM dataset, allowed the definition of more and better predictors with physical meaning and improved efficiency.

Keywords: seasonal precipitation, statistical forecast, Pampa plains, reanalysis data

Introduction

Northern Patagonia is a dry area with scarce precipitation which has been decreasing during the last 50 years which presents a major soil erosion problem. The agricultural and livestock activities are influenced by the interannual precipitation variability and therefore, crop yields are highly affected. The rainfall variability in Patagonia and eastern Los Andes has been studied by several authors. Castañeda and Gonzalez (2008) and Minetti et al. (2003) observed positive annual rainfall trends in northern and southern Patagonia and a decrease in the western and central areas in the last century using linear and non-linear techniques. Russian et al. (2010) and Aravena and Luckman (2008) studied the relation between the rainfall interannual variability in Patagonia and El Niño- Southern Oscillation (ENSO) and the Southern annular mode. Gonzalez (2013) investigated the rainfall and circulation connection which showed that those cyclonic anomalies, both over the surrounding Atlantic and Pacific Oceans and over the continent, enhance precipitation all over Patagonia all year long. Gonzalez and Herrera (2014) applied several statistical models to predict rainfall in Patagonia and the main result was that the consideration of an autoregressive integrated moving average model and Holt Winter in the multimodel ensemble improved the efficiency of the forecast. The physical basis of the seasonal climate predictability lies in the fact that slow variations in the boundary conditions of the earth (i.e., the sea surface temperature or soil wetness) can influence the atmospheric circulation and therefore other meteorological variables. To address this issue, dynamical and statistical models are derived. Therefore, the technology of numerical weather prediction is extended to longer scales and produces numerical climate predictions using state-of-the-art of global atmospheric models. The statistical forecasts are based on the detection of the relation between rainfall and circulation patterns previously observed . However, there is still a great deal of uncertainty about the efficiency of models in seasonal time scales. Many authors have pointed out the difficulties still detected when forecasting seasonal climate (Barnston et al. 2005; Leetmaa 2003; Coelho et al. 2005; Kumar 2006) and specially in South America where an evaluation of seasonal climate forecast has been done (Goddard 2003, Nobre et al. 2005; Barreiro 2009).

The efficiency of the different modes is due to several factors. It is important to point out that the meteorological variables have a random component impossible to predict but there are other different sources of error that can be addressed. One of them is the methodology used to build the forecast model and the other one is the data used to determine the best predictors to be included in the model. In this study statistical regression models were derived for a location in Buenos Aires plains for the summer season. The same method was used tu build the models using predictors defined with data from two different sets of reanalysis with the aim of detecting the error derived from that issue. Section 2 describes the data and methods while section 3 deals with the main results.

METHODS AND DATA

The aim of this paper is to build a statistical model to forecast summer (December-January-February) rainfall in Ascasubi station, located in

Buenos Aires plains in Argentina (39,4°S; 62,6°W) (Figure 1). Data has been provided by the National Meteorological Service of Argentina for the period 1979 - 2015. This is a critical area because agriculture is the most important activity and it is highly dependent on precipitation. The mean monthly rainfall evolution shows that precipitation is greater in summer than in winter (Figure 2). Although a decrease of annual and summer rainfall has been observed in the long term trends, it is not statistically significant (Figure 3). Two different sets of data were used to define the predictors used to build the models: ERA-INTERIM and NCEP reanalysis. ERA-INTERIM is a global atmospheric reanalysis from 1979, continuously updated in real time.

Figure 1. Location of the meteorological station used in the study.

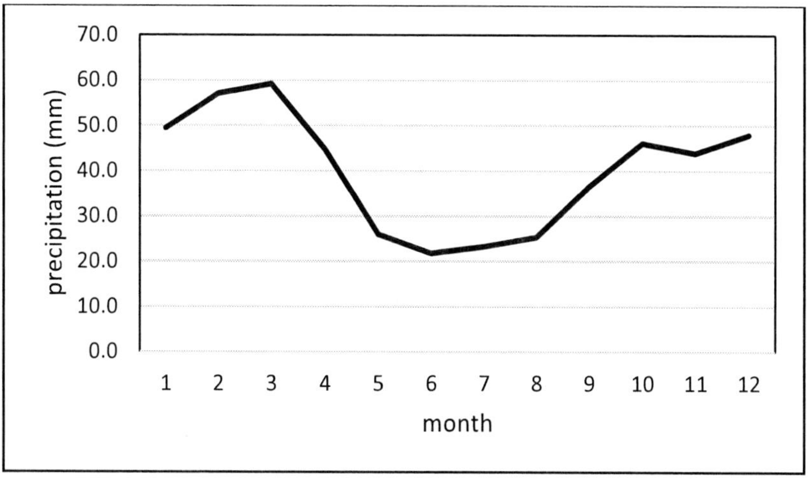

Figure 2. The mean monthly evolution of precipitation in ASCASUBI.

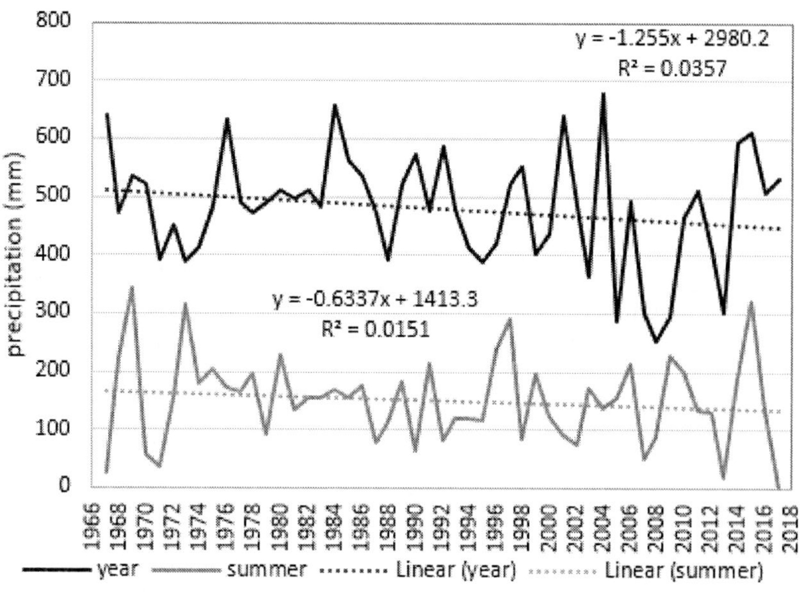

Figure 3. Long term trends for the annual (top) and summer (bottom) precipitation in ASCASUBI.

The spatial resolution of the data set is approximately 80km. The limitation is that ERA-INTERIM products are normally updated once per month, with a delay of two months, therefore they are not available for near forecast. The National Center for Atmospheric Prediction (NCEP) has developed a re-analysis project which generates global data sets for a long time period for different atmospheric variables. These reanalysis are derived using measured data from different sources, including observations from weather stations, ships, aircrafts, radiosondes and satellites. This data has a spatial resolution of 2,5° (approximately 220km).

The predictors have been defined using a correlation between summer rainfall and different oceanic and atmospheric variables in the previous month (November). The variables used are sea surface temperatures (SST), 500 HPa (G500), 1000 Hpa (G1000) and 200Hpa (G200) geopotential heights, zonal (U) and meridional (V) winds at 850 Hpa and precipitable water (PW) over the whole column of the atmosphere. Monthly anomalies were determined removing the climatological monthly means from the original values. Predictors are carefully selected in order to have physical meaning and to ensure they are independent from each other so as to avoid multicolineality problems (Wilks 1995). All the predictors were defined considering the areas where the correlation between summer rainfall and meteorological variables in November are statistically significant (95% confidence level).

Both datasets were used in order to determine the predictors to be used in the statistical models. The models were derived using lineal backward regression techniques (Hyndman and Athanasopoulos 2013). The best model for each data set was selected by considering the ten models which explain the major summer precipitation variance proportion and the one with the minimum cross-validation coefficient (CV) among them. The cross validation (Wilks 1995) is a method where n-1 years were used for calibration and the remaining year to validate the model. This process was repeated several times taking a different year as the validation target in each case. This method is generally strong in the presence of long-term climate variability and it is used specially when the amount of data is not so large. In the process of cross-validation, if the exclusion of a single year from the data does not cause important changes in the model there is no evidence of numerical instability. CV coefficient is defined as the mean square error of the differences between the observed and the predicted values. Under this criterion, the best model is the one with the smallest value of CV.

RESULTS

Using the criterion detailed in the above section, the best model for summer precipitation (SPn) derived from predictors from NCEP (Pn) data is:

$$SPn = 4438{,}42 + 24{,}22\ Pn1 - 0{,}66\ Pn2 - 7{,}47\ Pn3 \qquad (1)$$

where Pn1 is SST in the Pacific ocean related to El Niño, Pn2 is G500 over the South Pacific ocean and Pn3 is low level U over the Atlantic ocean (Figure 4). This model explains 44% of the summer precipitation variance in Ascasubi.

The best model for summer precipitation (SPe) using ERA-INTERIM datasets predictors (Pe) is:

$$SPe=-9869{,}56-0{,}2Pe1-19{,}27Pe2+60{,}53Pe3+30{,}21Pe4+16{,}9Pe5 \qquad (2)$$

where Pe1 is G500 in the South Pacific ocean, Pe2 is SST over the ENSO area, Pe3 SST over the Atlantic ocean, Pe4 SST over the Indian ocean and Pe5 is the low level U over the Pacfic ocean (Figure 4). This model explains 53,4% of the precipitation variance, a little more than in the case of NCEP data. It can be seen that using the same criteria the technique did not choose the same predictors to build the best model. A greater detail is observed in the selection predictors with the ERA-INTERIM set compared to those derived from the NCEP dataset. Figure 5 shows the observed and the predicted summer precipitation using both models (SPn and SPe) for Ascasubi. The adjustment is good, the correlation coefficient between observed rainfall and SPe is 0,77 and with SPn it is 0,63, both statistically significant (95% confidence level) but a slightly better for SPe than for SPn.

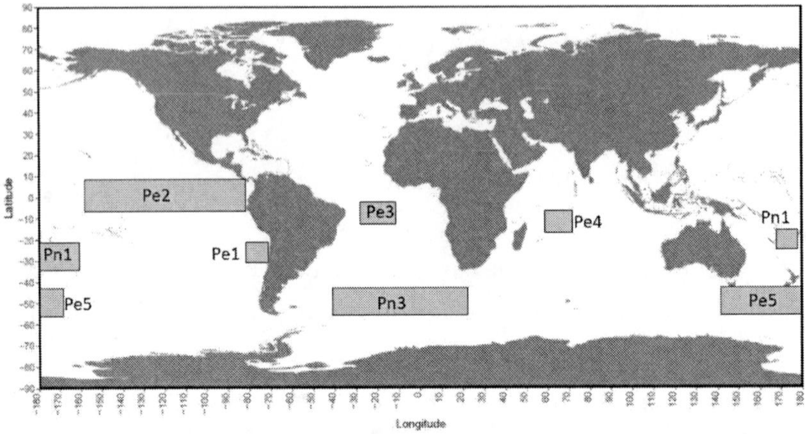

Figure 4. Location of the predictors used in the models.

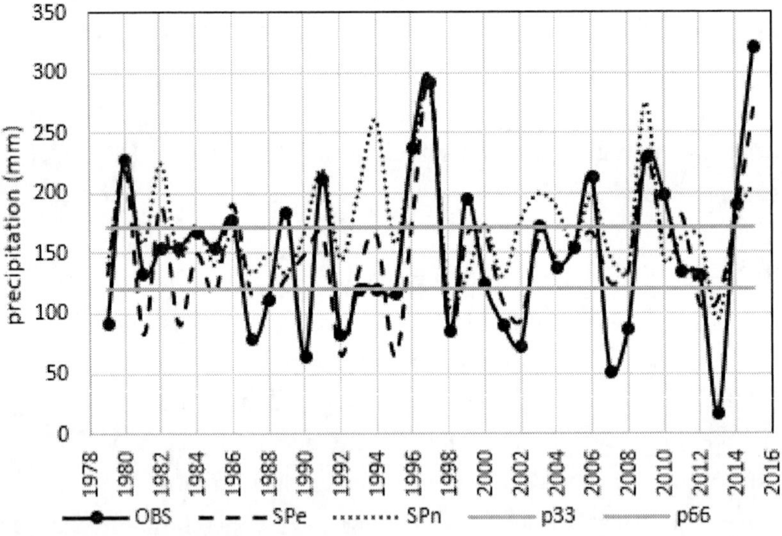

Figure 5. Observed (full line) and predicted summer precipitation with the model using ERA-INTERIM (SPe) and NCEP(SPn) datasets. The first and second terciles of observed precipitation are shown with gray horizontal lines.

Other metrics were computed to describe the efficiency of the models (Hyndman and Athanasopoulos, 2013). One of them was the Akaike's Information Criterion (AIC) defined as:

$$AIC = N \log(SSE/N) + 2(k+2) \tag{3}$$

where N is the number of observations used for estimation, k is the number of predictors in the model and SSE is the sum of square errors. The idea here is to penalise the fit of the model (SSE) with the number of parameters that need to be estimated. The model with the minimum value of the AIC is often the best model for forecasting. For large values of N, minimising the AIC is equivalent to minimising the CV value.

Also, the Schwarz's Bayesian Information Criterion (BIC) was computed, defined as:

$$BIC = T \log(SSE/N) + (k+2) \log(N) \quad (4)$$

BIC penalises the number of parameters more strongly than the AIC

While the percentage of explained variance (R^2) is the most widely used, sometimes as it selects too many predictor variables it becomes less suitable for forecasting. BIC has the feature that if there is a true underlying model, it will select that model provided enough data is given. Therefore, BIC, AIC or CV statistics can be used since each of them has forecasting as their objective. If the value of N is large enough, they will all lead to the same model.

Table 1 shows these coefficients for SPe and SPn. CV are very similar but AIC and BIC are better when using ERA-INTERIM predictors rather than NCEP ones.

The effectiveness of the different models was measured in semi-quantitative terms. The observed and predicted values were classified in three categories: "above normal" (greater than the second observed tercile), "normal" (between the first and the second observed terciles) and "below normal" (lower than the first observed tercile).

Table 1. Coefficient efficiency for the derived statistical forecast models using ERA-INTERIM (SPe) and NCEP (SPn) data

	CV	AIC	BIC	R^2
best model SPe	2534,03	289,41	300,69	0,534
best model SPn	2524,88	308,51	316,83	0,44

For each category the contingency table was calculated collapsing the 3x3 contingency table into three 2x2 tables. Each one is constructed by considering the “forecast event” as opposed to the complementary “non forecast event”.

Some measures of accuracy (Wilks 1995) were calculated for each category (Table 2). The hit rate (H) or the right proportion is the fraction of all the cases when the categorical forecast correctly anticipated the subsequent event. The probability of detection (POD) is defined as the fraction of those occasions when the forecasted event occurred on the one which was also forecast. The especificity (E) measures the proportion of actual negatives that are correctly identified as such. A receiver operating characteristic curve (ROC curve) is a graphical plot that illustrates the diagnostic ability of a binary classifier. The ROC curve is created by plotting POD against (1-E).

Table 2. Measures of accuracy calculated for each category and for both models (SRe and SPn)

	H	POD	1-E	AUC
below normal ERA	0,78	0,67	0,16	0,75
below normal NCEP	0,74	0,23	0,00	0,62
normal ERA	0,57	0,50	0,40	0,55
normal NCEP	0,54	0,54	0,46	0,54
above normal ERA	0,78	0,54	0,08	0,73
above normal NCEP	0,69	0,69	0,31	0,69

It can also be thought of as a plot of the Power as a function of the Type I Error of the decision rule. For each model it is a single point in the (POD, 1-E) space. The ROC curve is drawn by joining the axis of coordinates, the point (POD, 1-E) of the model and the point (1,1). The area under the curve (AUC) represents a meassure of efficiency and the model is better when this area is closer to 1. ROC analysis provides tools to select possibly optimal models and to discard suboptimal ones.

CONCLUSION

Results pointed out the best resolution and the methodology to build the reanalysis in ERA-INTERIM dataset, allowed the definition of more and better predictors with physical meaning and slightly improved efficiency. The summer precipitation variance explained by the ERA-INTERIM model and the AIC and BIC coefficients were greater than in the NCEP case. When categorized data was considered the ERA-INTERIM model seemed to have a slighly better hit rate than AUC coefficients but this was not observed in POD.

Although this paper describes the differences in the efficiency of the models in a single station, it can be pointed out that the uses of ERA-INTERIM dataset slightly improved the results.

REFERENCES

Aravena, J. C. and Luckman, B. H. 2008, Spatio-temporal rainfall patters in Southern South America, *Int. J. Climatol.,* DOI: 10.1002/joc 1761.

Barreiro, M. 2009. Influence of ENSO and the South Atlantic Ocean on climate predictability over Southeastern South America. *Climate Dynamics,* DOI 10.1007/s00382-009-0666-9.

Barnston, A., Kumar, A., Goddard, L. and Hoerling, M. 2005. Improving seasonal prediction practices thorough attribution of climate variability. *BAMS*, 59 - 72.

Castañeda, M. E. and González, M. H. 2008. Some aspects related to precipitation variability in the Patagonia region in Southern South America. *Atmósfera*, 21(3), 303 - 317.

Coelho, C., Stephenson, D., Balmaseda, M, Doblas Reyes, F. And Oldenborge, G. 2005. Towards an integrated seasonal forecasting system for South America. *J. Climate*, 19: 3704 - 3721.

Goddard, L., Barnston, A. and Mason, S. 2003. Evaluation of the IRI´s "net assesment" seasonal climate forecasts 1997-2001. *BAMS*, 1761 - 1781.

González, M. H. 2013. Some indicators of interannual rainfall variability in Patagonia (Argentine), in *Climate Variability - Regional and Thematic Patterns*, Chapter 6, Ed. Aondover Tarhule. INTECH, Rijeka, Croatia, pp. 133 - 161.

González, M. H. and Herrera, N. 2014. Statistical prediction of Winter rainfall in Patagonia (Argentina), in *Horizons in Earth Science Research,* vol.11, Chapter 7. 221-238. Ed. Benjamin Veress and Jozsi Szigethy, NOVA Publisher, NY, USA. http://dx.doi.org/10.5772/ 55494.

Hyndman, R. J.; Athanasopoulos, G. 2013. *Forecasting: principles and practice*. O Texts. 292 pages.

Kumar, A. 2006. On the interpretation and utility of skill information for seasonal climate predictions. *Mon.Wea. Rev.,* 135: 1974 - 1984.

Leetmaa, A. 2003. Seasonal Forecasting, Innovation in practice and institutions. *BAMS,* 84: 1686 - 1691.

Minetti, J. L.; Vargas, W. M.; Poblete, A. G.; Acuña, L. R. and Casagrande, G. 2003. Non-linear trends and low frequency oscillations in annual precipitation over Argentina and Chile, 1931-1999. *Atmósfera,* 16, 119 - 135. Nobre, C., Marengo, J., Cavalcanti, I., Obregon, G., Barros, V., Camilloni, I., Campos, N. and Ferreira, A. 2005. Seasonal to decadal predictability and prediction of South America Climate. *J. Climate*, 19, 23: 5988 - 6004.

Russian, G.; Agosta, E. and Compagnucci, R. H. 2010.Variabilidad interanual e interdecádica de la precipitación en la Patagonia norte. *Geoacta*, 35, 27 - 43.

Wilks, D. S. *Statistical methods in the atmospheric sciences (An introduction)*; International Geophysics series; Academic Press, San Diego, California, USA, 1995, 467 pp.

In: Agricultural Research Updates. Volume 27 ISBN: 978-1-53615-916-5
Editors: Prathamesh Gorawala et al.

Chapter 5

THEORETICAL FOURIER SERIES SOLUTION FOR HEAT CONDUCTION IN STONE FRUITS WITH INTERNAL HEAT SOURCE LINEARLY RELIANT ON TEMPERATURE

Francisco Javier Cuesta and María Dolores Alvarez*
Department of Characterization, Quality and Safety,
Institute of Food Science,
Technology and Nutrition (ICTAN-CSIC),
Ciudad Universitaria, Madrid, Spain

ABSTRACT

Analytical models for predicting postharvest cooling times at the stone–pulp interface in stone fruit, or for measuring the thermo-physical parameters of the pulp, mostly consider the pulp as a passive object for purposes of heat conduction, without internal heat generation or with a constant heat generation assigned. However, this is not strictly true. Like all other fruits, vegetables, etc., stone fruits respire, generating a heat that

* Corresponding Author's E-mail: mayoyes@ictan.csic.es.

increases with temperature and that should also be taken into account in equations. This chapter presents an analytical Fourier series solution to the equation for heat transfer by conduction in a spherical shell with an internal stone consisting of insulating material and considering an internal heat source linearly reliant on temperature, as a model of conduction of heat in stone fruits.

Keywords: stone fruit, transient heat transfer, Fourier series, heat of respiration

Nomenclature

A_0	Constant of the heat source (J m^{-3} s^{-1})
A_1	Slope of the heat source (J m^{-3} s^{-1} K^{-1})
$a = \frac{k}{\rho \cdot c}$	Thermal diffusivity (m^2 s^{-1})
$B_0 = A_0 + A_1 T_{ex}$	(Jm^{-3} s^{-1})
$Bi = \frac{hR}{k}$	Biot number
c	Thermal capacity per unit of mass (J s^{-1} kg^{-1} K^{-1})
$Fo = \frac{a \cdot t}{R^2}$	Fourier number (dimensionless time)
Fo_M	Dimensionless time at which the temperature profile reaches the maximum value
h	Surface heat transfer coefficient (J s^{-1} m^{-2} K^{-1})
J_i	Series expansion constants (dimensionless)
$J_{i,\infty}$	J_i when $Bi \to \infty$
$J_{i,0}$	Expansion constants for the no heat of respiration case
$J_{i,0,\infty}$	$J_{i,0}$ when $Bi \to \infty$
k	Thermal conductivity of the pulp (J s^{-1} m^{-1} K^{-1})
M, N	Intermediate constants
P	Constant appearing in Eq. (11) (dimensionless)
R	Radius of the fruit (m)
r	Radius from the centre of the fruit (m)
r_0	Radius of the stone (m)

S	Surface (m^2)
t	Time (s)
T	Temperature (°C)
T_{ex}	Temperature of the medium (°C)
T_0	Initial temperature of the sphere
T_{sf}	Temperature of the pulp at the external surface (°C)
$\Delta T = (T - T_{ex})$	Temperature difference between the fruit and the exterior (°C)
$\Delta T_0 = (T_0 - T_{ex})$	Initial temperature difference between the fruit and the exterior (°C)
V	Volume (m^3)
$x = r/R$	Dimensionless distance from the centre
$x_0 = r_0/R$	x at the stone–pulp interface (dimensionless)
$Y = (\theta - \theta_s)$	Dimensionless temperature difference

Greek Symbols

$\alpha^2 = \frac{A_1 R^2}{k}$	Dimensionless slope of the heat source
$\beta = \frac{B_0 R^2}{k \cdot \Delta T_0}$	Dimensionless number of the heat source
δ_n	Eigenvalues of the boundary Eq. (12) (dimensionless)
$\delta_{n,M} = \delta_n$	for $Bi \to \infty$ (dimensionless)
$\delta = \delta_1$	First eigenvalue of the boundary Eq. (12) (dimensionless)
$\theta = \frac{\Delta T}{\Delta T_0} = \frac{T - T_{ex}}{T_0 - T_{ex}}$	Dimensionless temperature difference
θ_M	Maximum value of θ
θ_S	Steady-state dimensionless difference of temperatures
θ_{Sf}	Dimensionless temperature difference at the external surface

$\bar{\theta}, \bar{\theta}_S$	Mass average θ and θ_S
ρ	Density of the pulp (kg m^{-3})
$\varphi(t)$	Function of time in the separation of variables
$\psi = \psi(\delta_n x)$	The spatial component of the solution in Fourier series expansion
$\psi' = \frac{\partial\psi(\delta x)}{\partial(\delta x)} = \frac{1}{\delta}\frac{\partial\psi(\delta x)}{\partial(x)}$	Derivative of the spatial function ψ
ψ_S	Steady-state spatial function ψ

INTRODUCTION

It is well known that in the design of postharvest processing of plant foods there are fundamental process parameters that must be controlled precisely, such as the cooling time needed to reach a certain temperature at the thermal centre of the product, the average temperature of the product at a given time, the surface heat transfer coefficient, etc. [1-4]. In most cases, since 1965, the calculation of these parameters is based on the first approximation to the general Fourier series solution for regular geometries in homogeneous isotropic bodies, which usually does not explicitly include the stone [5-14]. Nevertheless, stone fruits (olives, cherries, plums, etc.) present more or less spherical or ellipsoid geometries, but also they contain the seed (a ligneous core) in their interior, and with a physical and thermophysical parameters very dissimilar from those of the edible part or pulp. This fact entails that the theoretical thermal problems in stone fruits are different from those of solid bodies. In addition, the contact surface between seed and pulp is the deepest position that can be achieved in the stone fruit, carrying out a "thermal centre" role, which in homogeneous solid bodies is performed by the "geometric centre."

The problem of composite spherical solids with infinite heat transfer coefficient is traditionally considered in the literature as an application of the Laplace transformation [15]. However, recent literature, especially food literature, contains little information that includes analytical solutions for heat transfer explicitly including the seed. In this context we must consider the work of Helal [16], who proposed an integral transform method for nonlinear heat-conduction problems in multi-layered spherical media. Recently, Cuesta and Alvarez [17] proposed a Fourier series model for heat conduction in stone fruits. In the field of drying kinetics, an analytical solution for enhancing the drying rate of various fruits has been proposed by several authors [18-23]. Apart from these analytical works, we must resort to numerical calculation, where da Silva et al. [24, 25], Hernández-Díaz et al. [26], Fabbri et al. [27], Nilnont et al. [28], Ruiz-López et al. [29, 30], Sabarez [31] and Uyar and Erdoğdu [32] have proposed algorithms developed to simulate heat and mass transfer by numerical calculation in semi-spherical-geometry products, prunes and plums.

Nevertheless, the pulp is mostly considered as a passive object for purposes of heat conduction, without internal heat generation or with a constant heat generation assigned. However, this is generally not so. Like all other fruits, vegetables, etc., stone fruits respire, generating a heat that increases with temperature and that must also be taken into account in equations. Respiration heat is generally considered to be an exponential function of temperature [33-35]. However, other models are also used for practical purposes. Dincer [36, 37], Meffert et al. [38] and Mendonça et al. [39] take a constant value for heat generation. Wang [40] considers that heat generation is a constant in theory, but in practice assumes that it is negligible compared to other thermal sources. Other authors have used a potential model [34, 35, 41], or even a function of time [33]. There are regression models, such as those of Kole and Prasad [42] (regression to a fourth-grade polynomial) or Rao et al. [43] (regression to a sigmoidal model). Exact analytical solutions have been derived for some particular cases, including heat generation. Previously, Jakob [44] considered the case of conduction with a linear heat source under steady-state conditions, whereas Carslaw and Jaeger [15] proposed a solution to the problem in transient conditions for an infinite slab. Cuesta and Lamúa [11] proposed a Fourier series solution to the heat conduction equation in regular solids within a heat generation model linearly dependent on temperature.

However, none of these solutions takes into account the presence of the seed.

Accordingly, the main objective of this chapter is to simulate the temperature-time profile for stone fruits (cherries, peaches, plums, etc.) whose shape is approximately spherical, explicitly considering the respiration heat of the pulp, which in this paper is deemed to be a linear function of temperature. For this purpose, and as a theoretical simulation, this chapter proposes the solution in Fourier series to the respiration heat conduction equation of stone fruits consisting of an external spherical shell with an internal heat source linearly reliant on temperature and of an internal spherical core considered thermally quasi-insulating during the time in which the process takes place.

THEORETICAL ANALYSIS

General Case

Imagine a fruit whose shape will be considered spherical with outer radius R and whose seed will be considered a sphere concentric with the previous one and radius $r_0 < R$ (Fig. 1). This core is considered quasi-insulating, for purposes of heat conduction, compared to the external heat transfer in the times that the process lasts. It will be assumed that the initial temperature of the sphere is uniform (T_0), and that the fruit is placed in sudden contact with a mechanically-stirred medium at a uniform temperature T_{ex}. The fruit is considered to have an internal heat source linearly dependent on temperature.

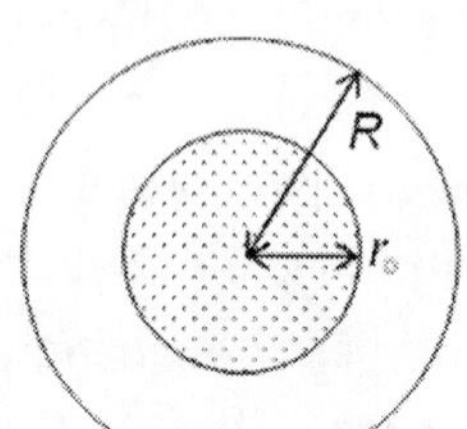

Figure 1. Esquematic of the model.

If k is the conductivity of the pulp (W m^{-1} K^{-1}), ρ its density (kg m^{-3}), c the thermal capacity per unit mass (J kg^{-1} K^{-1}), r the local radius (m), t the time (s) and h the surface heat transfer coefficient (Js^{-1} m^{-2} K^{-1}), according to Carslaw and Jaeger [15] and Gröber et al. [45] the one-dimensional Fourier equation for heat transfer may be written thus:

$$k\frac{1}{r^2}\frac{\partial}{\partial r}\left(r^2\frac{\partial T}{\partial r}\right) + A_0 + A_1 T = \rho c\frac{\partial T}{\partial t}$$

$$\mathrm{r} = \mathrm{R}:$$

$$\mathrm{k}\left[\frac{\partial \mathrm{T}}{\partial \mathrm{r}}\right]_{\mathrm{r=R}} = -\mathrm{h}(\mathrm{T_{Sf}} - \mathrm{T_{ex}})$$

T_{Sf} being the temperature of the pulp at the external surface.

The boundary condition at the stone–pulp interface is:

$$r = r_0:$$

$$\left[\frac{\partial T}{\partial r}\right]_{r=r_0} = 0$$

The above equations may be written in dimensionless form thus:

$$\frac{1}{x^2}\frac{\partial}{\partial x}\left(x^2\frac{\partial\theta}{\partial x}\right) + \alpha^2\theta + \beta = \frac{\partial\theta}{\partial Fo} \quad (1)$$

$$\left[\frac{\partial\theta}{\partial x}\right]_{x=1} = -Bi\cdot\theta_{Sf} \quad (2)$$

$$\left[\frac{\partial\theta}{\partial x}\right]_{x=x_0} = 0 \quad (3)$$

and, according to Cuesta and Lamúa [11], the heat source is expressed by two new dimensionless numbers:

$$\alpha^2 = \frac{A_1 R^2}{k} \quad (4)$$

Table 1. Summary of regression values calculated from ASHRAE® (2010)

Product	A_0 (W kg^{-1})	A_1 (W kg^{-1}K^{-1})
Apples "Y transparent"	0.0097	0.0073
Apples average	0.0061	0.0037
Apples early cultivars	0.0101	0.0040
Apples late cultivars	0.0052	0.0025
Apricots	0.0060	0.0038
Artichokes globe	0.0723	0.0124
Asparagus	0.1040	0.0478
Beans lima unshelled	0.0232	0.0211
Beans lima shelled	0.0303	0.0334
Beans snap	-0.0153	0.0189
Beets red roots	0.0160	0.0026
Blackberries	0.0251	0.0213
Blueberries	0.0028	0.0097
Broccoli	0.0291	0.0506
Brussels sprouts	0.0436	0.0179
Cabbage Penn State	0.0080	0.0048
Cabbage red early	0.0168	0.0070
Cabbage savoy	0.0165	0.0182
Cabbage white spring	0.0235	0.0085
Cabbage white winter	0.0104	0.0045
Cauliflower	0.0284	0.0089
Cauliflower	0.0298	0.0117
Celery	0.0120	0.0060
Gooseberries	0.0214	0.0039
Peas green	0.0486	0.0400
Plums	0.0050	0.0027
Potatoes	0.0174	0.0019
Raspberries	-0.0028	0.0222
Strawberries	-0.0033	0.0213

$$\beta = \frac{B_0 R^2}{k \cdot \Delta T_0} \tag{5}$$

where B_0 is:

$$B_0 = A_0 + A_1 T_{ex} \tag{6}$$

Table 1 shows the values of the constant and slope of the heat source (A_0 and A_1, respectively) obtained by linear regression of the average values taken from ASHRAE® [46].

Eq. (1) has to be integrated with the external and internal stone–pulp interface boundary conditions (2) and (3), respectively. Denominating the steady function as θ_s, the general solution may be written as follows (see Appendices A and B):

$$\theta(x, Fo) = \theta_s(x) + \sum_{i=1}^{\infty} J_i \cdot \psi(\delta_i x) \cdot e^{-(\delta_i{}^2 - \alpha^2)Fo} \tag{7}$$

where

$$\psi(\delta_i x) = \frac{\sin[\delta_i(x-x_0)] + \delta_i x_0 \cos[\delta_i(x-x_0)]}{\delta_i x} \tag{8}$$

and (see Appendix B, Eqs. B-8, B-14 and B-9):

$$\theta_s(\alpha x) = P \cdot \psi_s(\alpha x) - \frac{\beta}{\alpha^2} \tag{9}$$

$$\psi_s(\alpha x) = \frac{\sin[\alpha(x-x_0)]}{\alpha \cdot x} + \frac{x_0 \cdot \cos[\alpha(x-x_0)]}{x} \tag{10}$$

and

$$P = \frac{Bi \cdot \beta}{\alpha^2\left(\alpha \psi'_{s,\alpha} + Bi \cdot \psi_{s,\alpha}\right)} \tag{11}$$

The constants δ_i are the eigenvalues of the boundary equation (see Appendix A, Eq. (A-22)):

$$\delta = \tan[\delta(1 - x_0)] \frac{1 - Bi + \delta^2 x_0}{1 + x_0(Bi - 1)} \tag{12}$$

and the expansion constants J_i are (see Appendix C, Eqs. (C-20), (C-21)):

$$J_i = J_{i,0}\left(1 - \frac{\beta}{\delta_i^2 - \alpha^2}\right) \quad (13)$$

with

$$J_{i,0} = \frac{2Bi\psi(\delta_i)}{[\psi(\delta_i)]^2(\delta_i^2 + Bi^2 - Bi) - \delta_i^2 x_0^3} \quad (14)$$

In this chapter, the constants J_i (Eq. 13) are formally identical to those obtained in Cuesta and Lamúa [11]. However, the constants δ_i and $J_{i,0}$, as well as the functions $\psi(\delta_i x)$ and $\psi_s(\alpha x)$, are calculated according to Cuesta and Alvarez [17]. This formal structure will be maintained in the following sections.

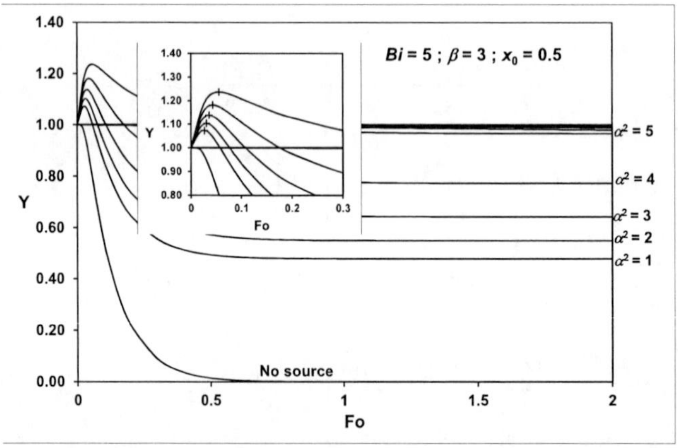

Figure 2. Temperature histories for $x_0 = 0.5$, $Bi = 5$, $\beta = 3$ and different values of α^2. In magnification, the crosses represent the approximation to the maximum values (θ_M) obtained by Eqs. (26) and (27) for the temperature histories depicted.

By way of illustration, Fig. 2 shows the influence of the parameter α^2 by plotting the temperature histories at the thermal centre (stone–pulp contact) for a fixed x_0 ($x_0 = 0.5$), Biot number ($Bi = 5$) and β ($\beta = 3$), and for different values of α^2 ($\alpha^2 = 1$ to 5). Fig. 3 shows the influence of the parameter β by plotting the temperature histories at the thermal centre (stone–pulp contact) for a fixed x_0 ($x_0 = 0.5$), Biot number ($Bi = 5$) and α^2 ($\alpha^2 = 2$), and for different values of β ($\beta = 1$ to 5). In turn, Fig. 4 shows the influence of the parameter x_0 by plotting the temperature histories at the thermal centre (stone–pulp contact) for a fixed Biot number ($Bi = 5$), α^2 ($\alpha^2 = 2$) and β ($\beta = 3$), and for different values of x_0 ($x_0 =$ 0.3 to 0.6).

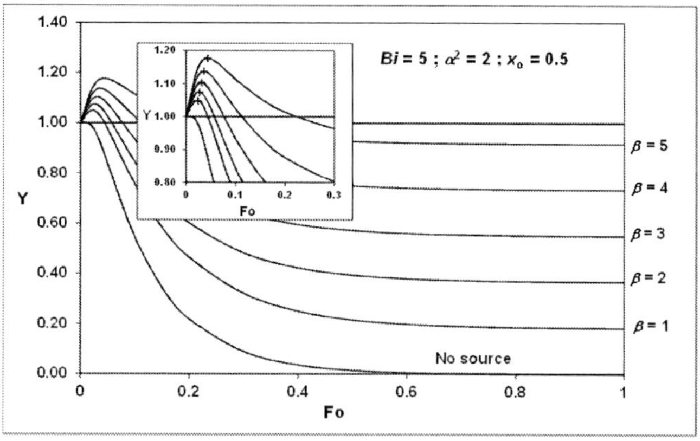

Figure 3. Temperature histories for $x_0 = 0.5$, $Bi = 5$, $\alpha^2 = 2$ and different values of β. In magnification, the crosses represent the approximation to the maximum values (θ_M) obtained by Eqs. (26) and (27) for the temperature histories depicted.

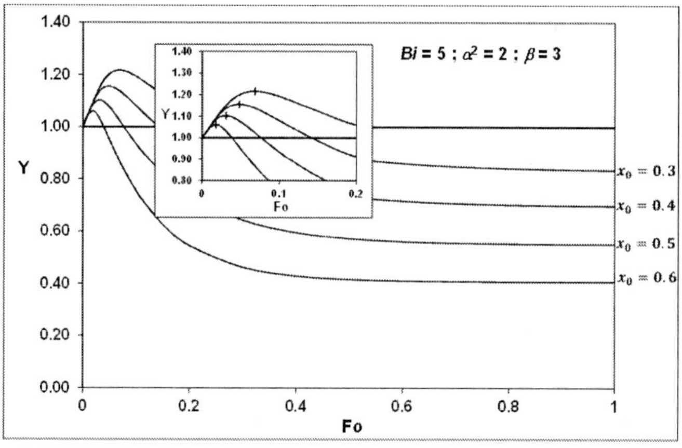

Figure 4. Temperature histories for $Bi = 5$, $\alpha^2 = 2$, $\beta = 3$ and different values of x_0. In magnification, the crosses represent the approximation to the maximum values (θ_M) obtained by Eqs. (26) and (27) for the temperature histories depicted.

General Case, $Bi \to \infty$

When $Bi \to \infty$ Eqs. (12) and (13) become:

$$\tan[\delta_M(1 - x_0)] + \delta_M x_0 = 0 \quad (15)$$

$$J_{i,\infty} = J_{i,0,\infty}\left(1 - \frac{\beta}{{\delta_{i,M}}^2 - \alpha^2}\right) \quad (16)$$

Where $J_{i,0,\infty}$ is:

$$J_{i,0,\infty} = (-1)^{(i-1)} \frac{2\sqrt{1 + {x_0}^2 {\delta_{i,M}}^2}}{1 + {x_0}^2 {\delta_{i,M}}^2 - {\delta_{i,M}}^2 {x_0}^3} \quad (17)$$

In the case $Bi \to \infty$ is identified by the subscript ∞ in coefficients J_i. In the case of the eigenvalues δ_n the case $Bi \to \infty$ will be identified by the subscript M ($\delta_{n,M}$) to make explicit that it is the maximum value of δ_n attainable in the particular pulp-to-stone configuration. The reason for this is that, in practice, the first eigenvalue (δ_1) of the transcendental boundary Eq. (12) is one of the most important parameters in the process, as $-{\delta_1}^2$ is the absolute slope of the linear portion of the cooling/heating curve in semi-logarithmic scale (${\delta_1}^2 = -d(\ln Y)/d(Fo)$), and knowledge of it allows us to determine the surface heat transfer coefficient. Also, $\delta_{1,M}^2$ is the maximum slope achievable, and it only depends on the shape of the object. In our case, because the geometry is spherical, it depends only on the size of the stone. Furthermore, Eqs. (12) to (17) reduce to those of the solid sphere when $x_0 = 0$.

Mass Average

As the density is considered to be constant, the mass average temperature is:

$$\bar{\theta}(Fo) \equiv \bar{\theta} = \frac{1}{V}\int_{x_0}^{1} \theta(x, Fo)dV = \bar{\theta}_s + \sum_{i=1}^{\infty} \bar{J}_i \cdot e^{-(\delta_i{}^2-\alpha^2)Fo} \tag{18}$$

where, as shown in Appendices B and C (Eqs. (B-30) and (C-32)):

$$\bar{\theta}_s = -\frac{\beta}{\alpha^2}\left[\frac{3}{\alpha\cdot(1-x_0{}^3)}\cdot\frac{Bi\cdot\psi'_{s,\alpha}}{\alpha\psi'_{s,\alpha}+Bi\cdot\psi_{s,\alpha}} + 1\right] \tag{19}$$

and:

$$\bar{J}_i = \frac{6Bi^2}{\delta_i{}^2(1-x_0{}^3)}\frac{\psi^2(\delta_i)}{[\psi(\delta_i)]^2(\delta_i{}^2+Bi^2-Bi)-\delta_i{}^2x_0{}^3}\left(1-\frac{\beta}{\delta_i{}^2-\alpha^2}\right) \tag{20}$$

and when $Bi \to \infty$ (Eq. B-31):

$$\bar{\theta}_{s,\infty} = -\frac{\beta}{\alpha^2}\left[\frac{3}{\alpha\cdot(1-x_0{}^3)}\cdot\frac{\psi'_{s,\alpha}}{\psi_{s,\alpha}} + 1\right] \tag{21}$$

and

$$\bar{J}_{i,\infty} = \frac{6}{\delta_{i,M}{}^2(1-x_0{}^3)}\cdot\frac{1+\delta_{i,M}{}^2x_0{}^2}{1+\delta_{i,M}{}^2x_0{}^2-\delta_{i,M}{}^2x_0{}^3}\cdot\left(1-\frac{\beta}{\delta_{i,M}{}^2-\alpha^2}\right) \tag{22}$$

APPROXIMATION TO MAXIMUM VALUE AT THE PULP-TO-STONE INTERFACE

As we can see in Figs. 2–4, depending on the particular conditions in each case, the temperature may rise at the outset up to a given maximum value θ_M in a time Fo_M, thereafter falling to its steady value. This value can be approximated by Eq. (D-14) (see Appendix D):

$$Fo_M \approx \frac{\ln\left[-\frac{(\delta_2{}^2-\alpha^2)\cdot J_2}{(\delta_1{}^2-\alpha^2)\cdot J_1}\right]}{(\delta_2{}^2-\delta_1{}^2)} \tag{23}$$

To which corresponds the value:

$$\theta_M \approx \theta_{s,x_0} + J_1 \cdot e^{-({\delta_1}^2-\alpha^2)Fo_M} + J_2 \cdot e^{-({\delta_2}^2-\alpha^2)Fo_M} \quad (24)$$

Figs. 2–4 also show the maximum values of the dimensionless temperature θ_M obtained from Eqs. (23) and (24).

Threshold Biot Number

For the product to cool, the rate of extraction of heat via the surface must be greater than the heat generated in the interior. This can be expressed by the following inequality (see Appendix B; Eq. (B-23)):

$$\alpha^2 < \delta^2 \leq {\delta_M}^2 \quad (25)$$

Otherwise, following the reasoning of Jakob [44] and Carslaw and Jaeger [15], the hypothetical case $\alpha^2 > \delta^2$ would mean that, with that value of the surface heat transfer coefficient, the extraction of heat via the surface would not be sufficient to eliminate the heat generated in the interior and the temperature rise would be unbounded.

Eq. (25) implies that the Biot number must be greater than a certain threshold Biot number (Eq. (B-21)):

$$Bi \geq Bi_{Th} = -\frac{\alpha \cdot {\psi'}_{s,\alpha}}{\psi_s(\alpha)} \quad (26)$$

That is, the surface heat transfer coefficient must be greater than a certain threshold value:

$$h > h_{Th} = -\frac{k}{R}\frac{\alpha \cdot {\psi'}_{s,\alpha}}{\psi_s(\alpha)} \quad (27)$$

It must be remembered that:

$$\psi_{s,\alpha} = \frac{\sin[\alpha(1-x_0)]}{\alpha} + x_0 \cdot \cos[\alpha(1-x_0)] \tag{28}$$

$$\psi'_{s,\alpha} = \frac{\cos[\alpha(1-x_0)]}{\alpha}(1-x_0) - \frac{\sin[\alpha(1-x_0)]}{\alpha}\left(x_0\alpha + \frac{1}{\alpha}\right) \tag{29}$$

Therefore the threshold Biot number also depends on x_0.

Figure 5 shows the threshold Biot number vs. $\boldsymbol{\alpha^2}$ for $\boldsymbol{x_0 = 0 \text{ to } 0.6}$

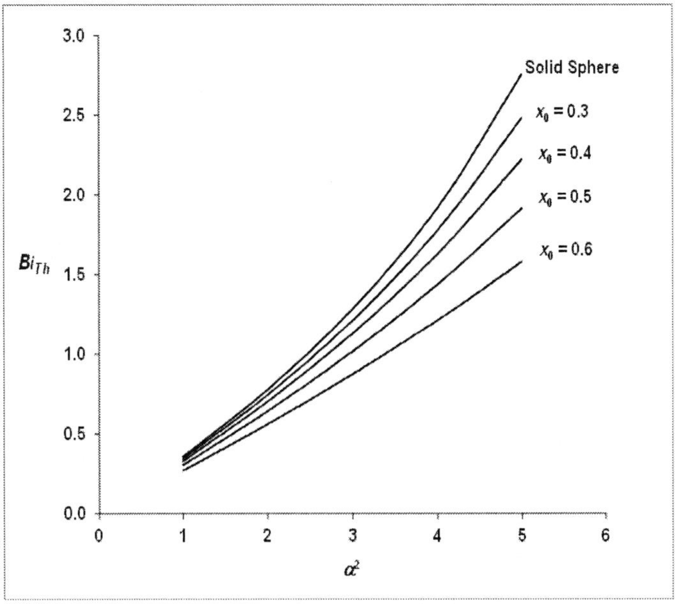

Figure 5. Threshold Biot number for the cases shown in Figures 2–4.

Illustrative Example

As can be seen in appendix D, one of the applications of the first approximations is the indirect determination of the thermal diffusivity and the surface heat transfer coefficient (section: *Indirect measurement of thermal diffusivity and surface heat transfer coefficient*). The condition is that the thermal conductivity and the two heat generation constants A_0 and A_1 are known.

A plum (*Prunus domestica*) weighing 125.77 g was stabilized (for 24 h) at $T_0 = 21.8$ °C and a 4-l bath was also stabilized into a storage chamber (for 24 h as well) whose set point temperature was 4.5 °C. At the time of introducing the plum into the bath, the water was stabilized at 4.6 °C, approximately.

The fruit approached a prolate spheroid with the larger diameter = 59.88 mm and the smaller diameter = 55.67 mm. After the complete experience, the stone was measured. Its dimensions were as follows: weight 2.060 g, larger diameter = 27.68 mm; intermediate diameter = 17.56 mm and smaller diameter = 11.11 mm.

The pulp weight is then $M = 60 - 0.6125 = 0.05939$ kg.

The external measured volume is:

$$V = \frac{4}{3}\pi R_1 R_2 R_3 = 104516 \times 10^{-4} \text{m}^3$$

So, the equivalent radius:

$$R = \sqrt[3]{R_1 R_2 R_3} = 0.02922 \text{ m}$$

The stone's volume $2.8275 \times 10^{-6} \text{m}^3$

The equivalent radius of the stone $r_0 = 0.00877$ m

The measured pulp density was therefore:

$$\rho = \frac{M_{tot} - M_{stone}}{V_{tot} - V_{stone}} = 1216.53 \text{ kg m}^{-3}$$

The ratio stone/fruit radius is:

$$x_0 = \frac{r_0}{R} = 0.30$$

The thermal conductivity for the plum is $k = 0.247 \text{ J/s} \cdot \text{m} \cdot \text{K}$ [46].

And from ASHRAE [46], the linear regression of the heat of respiration constants (average values) between 0 and 20 °C are:

$$A_0 = 5.290 \mathrm{J} \cdot \mathrm{m}^{-3} \cdot \mathrm{s}^{-1}$$

$$A_1 = 2.795 \mathrm{J} \cdot \mathrm{m}^{-3} \cdot \mathrm{s}^{-1} \cdot \mathrm{K}^{-1}$$

That is:

$$B_0 = A_0 + A_1 T_{ex} = 18.059 \mathrm{J} \cdot \mathrm{m}^{-3} \cdot \mathrm{s}^{-1}$$

$$\alpha^2 = \frac{A_1 R^2}{k} = 9.663 \times 10^{-3}$$

$$\beta = \frac{B_0 R^2}{k \cdot \Delta T_0} = 3.623 \times 10^{-3}$$

From the original time (s) – temperature (°C) tables, the average temperature of the water was $T_{ex} = 4.568$ °C and the average temperature of the plum at the stone/fruit interface was approximately 4.594 °C That is, $\theta_s = (T_s - T_{ex})/\Delta T_0 \approx 1.515 \times 10^{-3}$.

Otherwise, from the original time (s) – temperature (°C) tables, it is possible to derive the $\{t - Y\}$ table, which is shown in Fig. 6a on a semi-logarithmic scale, where the linear portion is clearly visible. The linear regression is:

$$\ln Y_c = M - S \cdot t \approx 0.3243 - 1.0255 \times 10^{-3} \cdot t$$

With a coefficient$R^2 = 0.999$.

That is:

$$J_1 = e^M = 1.3830$$

The slope δ^2 corresponding to this value J_1 regardless of the heat source is:

$$\delta^2 = 5.0304$$

Which will be considered as a starting point for the difference $(\delta^2 - \alpha^2)$.

So, substituting into Eq. (D-32):

$$J_0 = \frac{1.3830}{\left(1 - \frac{3.623\times10^{-3}}{5.0304}\right)} = 1.3840 = J_{0,0}$$

Then following steps 8 through 11 of the process described in section: *Indirect measurement of thermal diffusivity and surface heat transfer coefficient* of Appendix D, the following values are obtained:

$$\delta = 2.246$$

Corresponding to the Biot Number:

$$Bi = 2.517$$

Therefore, clearing a in Equation (D-28) we get to:

$$a = 1.740 \times 10^{-7} \mathrm{m^2 s^{-1}}$$

and, clearing h in Biot number:

$$h = 21.27\ \mathrm{Jm^{-2}s^{-1}K^{-1}}$$

As the Biot number is already known, the roots of the transcendental boundary equation (Eq. 12) and the constants of the serial expansion (Eqs. 13 and 14) can be calculated. So, the complete theoretical time-temperature curve of this experiment can be drawn and compared with the experimental values. The standard error between experimental and theoretical Y curves is ± 0.27 °C.

The result of the complete series corresponding to this Biot number is shown in Fig. 6b (unbroken line), along with the experimental values (empty circles).

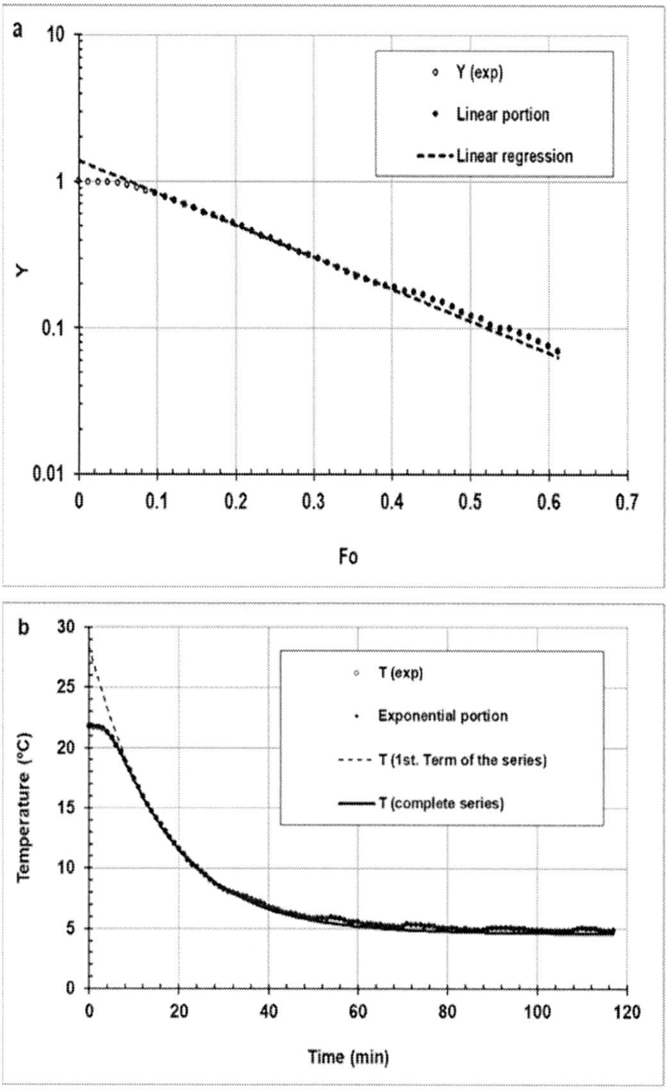

Figure 6. (a) Linear portion and linear regression (in semi logarithmic scale); (b) Experimental kinetic of temperature, exponential portion, first term of the series and complete series.

CONCLUSION

An analytical Fourier series solution to the equation for heat transfer by conduction in a spherical shell with an internal stone consisting of insulating material and considering an internal heat source linearly reliant on temperature has been proposed. The new constants of the Fourier series depend both on the ratio of the stone diameter to the fruit diameter and on the dimensionless intensity of the respiration heat number (α^2). This α^2 number plays a decisive role in the process, being the main dimensionless number of respiration of the fruit. The main consequence of α^2 is that the temperature initially increases until it reaches a certain maximum value θ_M in a time Fo_M, thereafter falling to its steady value. Two simple equations are proposed for approximate calculation of both the maximum temperature (θ_M) value and the time to attain it (Fo_M) at the pulp-to-stone interface. Another consequence of α^2 is the existence of a threshold value for the Biot number (Bi_{Th}), the value of which has also been determined in the present paper. Knowledge of these three values may be important in the design of post-harvest processing of plant foods.

APPENDICES A-D

Appendix A. General Solution

Eq. (A-1) is to be integrated:

$$\frac{1}{x^2}\frac{\partial}{\partial x}\left(x^2\frac{\partial\theta}{\partial x}\right) + \alpha^2\theta + \beta = \frac{\partial\theta}{\partial Fo} \tag{A-1}$$

with the external boundary condition (A-2):

$$\left[\frac{\partial\theta}{\partial x}\right]_{x=1} = -Bi\cdot\theta_{Sf} \tag{A-2}$$

and the internal stone-pulp boundary condition (A-3):

$$\left[\frac{\partial \theta}{\partial x}\right]_{x=x_0} = 0 \tag{A-3}$$

By definition the steady-state solution does not depend on time, and therefore the second term of Eq. (A-1) is cancelled when applied to that case. If we mark the steady-state function with subscript "*s*", we get:

$$\frac{1}{x^2}\frac{\partial}{\partial x}\left(x^2\frac{\partial \theta_s}{\partial x}\right) + \alpha^2\theta_s + \beta = \frac{\partial \theta_s}{\partial Fo} = 0 \tag{A-4}$$

$$\left[\frac{\partial \theta_s}{\partial x}\right]_{x=1} = -Bi \cdot (\theta_s)_{sf} \tag{A-5}$$

$$\left[\frac{\partial \theta_s}{\partial x}\right]_{x=x_0} = 0 \tag{A-6}$$

where:

$$\theta_s = \frac{T_s - T_{ex}}{T_0 - T_{ex}} \tag{A-7}$$

If we subtract Eqs. (A-4) to (A-6) from Eqs. (A-1) to (A-3), respectively, and denote, for convenience's sake:

$$Y = (\theta - \theta_s) \tag{A-8}$$

this leaves:

$$\frac{1}{x^2}\frac{\partial}{\partial x}\left[x^2\frac{\partial Y}{\partial x}\right] + \alpha^2 Y = \frac{\partial Y}{\partial Fo} \tag{A-9}$$

$$\left[\frac{\partial Y}{\partial x}\right]_{x=1} = -Bi \cdot Y_{sf} \tag{A-10}$$

$$\left[\frac{\partial Y}{\partial x}\right]_{x=x_0} = 0 \qquad \text{(A-11)}$$

Eq. (A-9) admits a solution in separation of variables in the same way as without a heat source. The complete solution will therefore be a series expansion, as can easily be proved. In fact, let Y be:

$$Y = \psi(x)\varphi(t)$$

Substituting into Eq. (A-9) and rearranging terms leads to:

$$\frac{1}{\psi(x)}\frac{1}{x^2}\frac{\partial}{\partial x}\left(x^2\frac{\partial\psi(x)}{\partial x}\right) = \frac{1}{\varphi(t)}\frac{\partial\varphi(t)}{\partial Fo} - \alpha^2 = -\delta^2$$

That is:

$$\frac{1}{\varphi(t)}\frac{\partial\varphi(t)}{\partial Fo} = -(\delta^2 - \alpha^2)$$

The solution of which is:

$$\varphi(t) = e^{-(\delta^2-\alpha^2)Fo} \qquad \text{(A-12)}$$

and

$$\frac{\partial}{\partial x}\left(x^2\frac{\partial\psi(x)}{\partial x}\right) = -\delta^2x^2\psi(x) \qquad \text{(A-13)}$$

The solution of which can be written as:

$$\psi(\delta x) = M\frac{\sin(\delta x)}{x} + N\frac{\cos(\delta x)}{x} \qquad \text{(A-14)}$$

In the case of the homogeneous solid sphere, the symmetry condition in $x = 0$ rules out the function of the second term, but in the present case such symmetry does not occur explicitly but in the context of a core having a radius $x_0 = r_0/R$. Since the derivative vanishes at $r = r_0$ for convenience's sake the solution explicitly marks the difference $(x - x_0)$. This is possible since all that changes is the value of the constants *M* and *N*. Thus, the solution can be written in the form:

$$\psi(\delta x) = M\frac{\sin[\delta(x-x_0)]}{x} + N\frac{\cos[\delta(x-x_0)]}{x} \tag{A-15}$$

This satisfies the differential Eq. (A-13), as can be verified by simply substituting and regrouping its terms.

$$\frac{\partial\psi(\delta x)}{\partial x} = \frac{\cos[\delta(x-x_0)]}{x}\left(M\delta - \frac{N}{x}\right) - \frac{\sin[\delta(x-x_0)]}{x}\left(N\delta + \frac{M}{x}\right) \tag{A-16}$$

The aim is for the pulp–stone contact to play the same role as is played by the thermal centre in the homogeneous solid sphere; therefore, we take *M* and *N* so that $\psi(\delta x_0) = 1$, which is possible since the constants of the complete solution are calculated afterwards.

The values M and N can be calculated by taking into account that at $x = x_0$, $\sin[\delta(x - x_0)] = 0$ and $\cos[\delta(x - x_0)] = 1$, and hence, substituting into (A-15):

$$N = x_0 \tag{A-17}$$

Substituting in Eq. (A- 15), taking into account condition (A-11) and resolving:

$$M = \frac{1}{\delta} \tag{A-18}$$

Thus, the function in general is left as:

$$\psi(\delta x) = \frac{\sin[\delta(x-x_0)] + \delta x_0 \cos[\delta(x-x_0)]}{\delta x} \tag{A-19}$$

Its derivative is:

$$\frac{\partial\psi(\delta x)}{\partial x}=\frac{\cos[\delta(x-x_0)]}{x}\left(1-\frac{x_0}{x}\right)-\frac{\sin[\delta(x-x_0)]}{x}\left(x_0\delta+\frac{1}{\delta x}\right) \quad \text{(A-20)}$$

Hereafter, as long as there is no risk of confusion, we will use the notation:

$$\psi=\psi(\delta x);\ \psi_\delta\equiv\psi(\delta)$$

$$\psi'=\frac{\partial\psi}{\partial(\delta x)}=\frac{1}{\delta}\frac{\partial\psi}{\partial(x)}\rightarrow\frac{\partial\psi(\delta x)}{\partial(x)}=\delta\psi'$$

$$\psi'_\delta\equiv\psi'(\delta)=\left[\frac{\partial\psi(\delta x)}{\partial(\delta x)}\right]_{x=1}=\frac{1}{\delta}\left[\frac{\partial\psi(\delta x)}{\partial(x)}\right]_{x=1};\ \left[\frac{\partial\psi(\delta x)}{\partial(x)}\right]_{x=1}=\delta\psi'_\delta$$

The constants δ are derived from the boundary Eq. (A-10):

$$\left[\frac{\partial Y}{\partial x}\right]_{x=1}=\left[\frac{\partial\psi(\delta x)}{\partial x}\right]_{x=1}=\delta\psi'_\delta=-Bi\cdot\psi_\delta \quad \text{(A-21)}$$

From Eqs. (A-19), (A-20) and (A-21), and by rearranging terms, the transcendental boundary equation is:

$$\delta=\tan[\delta(1-x_0)]\frac{1-Bi+\delta^2x_0}{1+x_0(Bi-1)} \quad \text{(A-22)}$$

As in the case of the homogeneous solid sphere, the number of possible solutions is infinite, and therefore the solution can be written in the form of a Fourier series:

$$Y(x,Fo)=\sum_{i=1}^{\infty}J_i\cdot\psi(\delta_i x)\cdot e^{-(\delta_i{}^2-\alpha^2)Fo} \quad \text{(A-23)}$$

That is, undoing the change indicated in Eq. (A-8):

$$\theta=\theta_s+Y=\theta_s+\sum_{i=1}^{\infty}J_i\cdot\psi(\delta_i x)\cdot e^{-(\delta_i{}^2-\alpha^2)Fo} \quad \text{(A-24)}$$

Eq. (A-24) is the solution with the following steady-state function θ_s (corresponding to $t = \infty$ (see Appendix B):

$$\theta_s(\alpha x) = P \cdot \psi_s(\alpha x) - \frac{\beta}{\alpha^2} \tag{A-25}$$

where:

$$\psi_s(\alpha x) = \frac{\sin[\alpha(x-x_0)]}{\alpha \cdot x} + \frac{x_0 \cdot \cos[\alpha(x-x_0)]}{x} \tag{A-26}$$

and:

$$P = \frac{Bi \cdot \beta}{\alpha^2\left[\alpha \cdot \psi'_{s,\alpha} + Bi \cdot \psi_{s,\alpha}\right]} \tag{A-27}$$

Coefficients J_i depend on the initial and boundary conditions imposed, and they are calculated in Appendix C:

$$J_i = J_{i,0}\left(1 + \frac{\beta}{\alpha^2 - {\delta_i}^2}\right) \tag{A-28}$$

$J_{i,0}$ being the J_i coefficient for the case without internal heat sources [17]:

$$J_{i,0} = \frac{2Bi\psi(\delta_i)}{[\psi(\delta_i)]^2({\delta_i}^2 + Bi^2 - Bi) - {\delta_i}^2 {x_0}^3} \tag{A-29}$$

When $Bi \to \infty$,

$$\theta_\infty = \theta_{s,\infty} + \sum_{i=1}^{\infty} J_{i,\infty} \cdot \psi(\delta_{i,M} x) \cdot e^{-({\delta_{i,M}}^2 - \alpha^2)Fo} \tag{A-30}$$

where (see Appendix B):

$$\theta_{s,\infty}(\alpha x) = \frac{\beta}{\alpha^2}\left[\frac{\psi_s(\alpha x)}{\psi_{s,\alpha}} - 1\right] \tag{A-31}$$

and (see Appendix C):

$$J_{i,\infty} = J_{i,0,\infty}\left(1 + \frac{\beta}{\alpha^2 - \delta_i^2}\right) \tag{A-32}$$

with:

$$J_{i,0,\infty} = \frac{2\sqrt{1+x_0^2\delta_{i,M}^2}}{1+x_0^2\delta_{i,M}^2 - \delta_{i,M}^2 x_0^3} \tag{A-33}$$

Otherwise, when $Bi \to \infty$, the left-hand side of Eq. (A-21) remains finite, because $\psi'_{\delta_{i,M}}$ is finite. Therefore: $\psi(\delta_{i,M}) = 0$. That is:

$$\sin[\delta_{i,M}(1 - x_0)] + \delta_{i,M}x_0\cos[\delta_{i,M}(1 - x_0)] = 0 \tag{A-34}$$

or:

$$\tan[\delta_{i,M}(1 - x_0)] + \delta_{i,M}x_0 = 0 \tag{A-35}$$

and, as can easily be deduced, the limit:

$$\lim_{Bi\to\infty}(Bi \cdot \psi_{\delta_{i,M}}) = -\delta_{i,M}\psi'_{\delta_M} = \mp\sqrt{1 + x_0^2\delta_{i,M}^2} \tag{A-36}$$

Mass Average

As the density is considered to be constant, the mass average temperature in the spherical shell of the pulp is:

$$\bar{\theta}(Fo) \equiv \bar{\theta} = \frac{1}{V}\int_{x_0}^{1} \theta(x, Fo)dV = \bar{\theta}_s + \sum_{i=1}^{\infty} \bar{J}_i \cdot e^{-(\delta_i^2 - \alpha^2)Fo} \tag{A-37}$$

V being the volume of the spherical shell of the pulp:

$$V = \frac{4}{3}\pi(R^3 - {r_0}^3) \tag{A-38}$$

$$dV = 4\pi r^2 dr \tag{A-39}$$

As can be seen in Appendices B and C, respectively:

$$\bar{\theta}_s = -\frac{\beta}{\alpha^2}\left[\frac{3}{\alpha\cdot(1-{x_0}^3)}\cdot\frac{Bi\cdot\psi'_{s,\alpha}}{\alpha\psi'_{s,\alpha}+Bi\cdot\psi_{s,\alpha}}+1\right] \tag{A-40}$$

and:

$$\bar{J}_i = \frac{6Bi^2}{{\delta_i}^2(1-{x_0}^3)}\frac{\psi^2(\delta_i)}{[\psi(\delta_i)]^2({\delta_i}^2+Bi^2-Bi)-{\delta_i}^2{x_0}^3}\left(1+\frac{\beta}{\alpha^2-{\delta_i}^2}\right) \tag{A-41}$$

Appendix B. Steady State

Steady-state Eq. (B-1) is to be integrated:

$$\frac{1}{x^2}\frac{\partial}{\partial x}\left(x^2\frac{\partial\theta_s}{\partial x}\right)+\alpha^2\theta_s+\beta=0 \tag{B-1}$$

$$\left[\frac{\partial\theta_s}{\partial x}\right]_{x=1} = -Bi\cdot(\theta_s)_{sf} \tag{B-2}$$

$$\left[\frac{\partial\theta_s}{\partial x}\right]_{x=x_0} = 0 \tag{B-3}$$

where:

$$\theta_s = \frac{T_s-T_{ex}}{T_0-T_{ex}} \tag{B-4}$$

A particular solution of Eq. (B-1) is:

$$\theta_p = -\frac{\beta}{\alpha^2} \tag{B-5}$$

and the homogeneous differential equation corresponding to Eq. (B-1) will be:

$$\frac{1}{x^2}\frac{\partial}{\partial x}\left(x^2\frac{\partial\theta_h}{\partial x}\right)+\alpha^2\theta_h=0 \tag{B-6}$$

This equation is formally identical to the one corresponding to the case with no heat source, except that here the coefficient α^2 is imposed by the product's internal heat source and therefore is unique:

$$\theta_h=\psi_s(\alpha x) \tag{B-7}$$

Consequently, the steady-state equation is (from Eqs. (B-5) and (B-7)):

$$\theta_s(\alpha x)=P\cdot\psi_s(\alpha x)-\frac{\beta}{\alpha^2} \tag{B-8}$$

and the boundary condition (B-2) leads to:

$$P=\frac{Bi\cdot\beta}{\alpha^2\left(\alpha\psi'_{s,\alpha}+Bi\cdot\psi_{s,\alpha}\right)} \tag{B-9}$$

Therefore, finally:

$$\theta_s(\alpha x)=\frac{\beta}{\alpha^2}\left(\frac{Bi\cdot\psi_s(\alpha x)}{\alpha\psi'_{s,\alpha}+Bi\cdot\psi_{s,\alpha}}-1\right) \tag{B-10}$$

As in the case of $\psi(\delta x)$ (see Appendix A), $\psi_s(\alpha x)$ may be written as:

$$\psi_s(\alpha x)=M_s\frac{\sin[\alpha(x-x_0)]}{x}+N_s\frac{\cos[\alpha(x-x_0)]}{x} \tag{B-11}$$

The condition $\psi_s(\alpha x_0)=1$ leads to:

$$N_s=x_0 \tag{B-12}$$

and the internal boundary condition (Eq. (B-3)) leads to:

$$M_s = \frac{1}{\alpha} \tag{B-13}$$

As a result:

$$\psi_s(\alpha x) = \frac{\sin[\alpha(x-x_0)]}{\alpha \cdot x} + \frac{x_0 \cdot \cos[\alpha(x-x_0)]}{x} \tag{B-14}$$

When $Bi \to \infty$:

$$P_\infty = \frac{Bi \cdot \beta}{\alpha^2\left(\alpha\psi'_{s,\alpha} + Bi \cdot \psi_{s,\alpha}\right)} = \frac{\beta}{\alpha^2} \cdot \frac{1}{\psi_{s,\alpha}} \tag{B-15}$$

and:

$$\theta_{s,\infty}(\alpha x) = \frac{\beta}{\alpha^2}\left[\frac{\psi_s(\alpha x)}{\psi_{s,\alpha}} - 1\right] \tag{B-16}$$

Existence Conditions and Threshold Biot Number

At this point there are two points to be clarified. The first one is that $\alpha = \sqrt{\alpha^2}$ is the dimensionless constant of the heat source intensity, and therefore the external boundary (Eq. (B-2)) is not an eigenvalue equation.

The second point is that, following the reasoning of Jakob [44], we can state that there are two conditions that Eq. (B-10) must meet to remain positive:

$$1)\ \alpha\psi'_{s,\alpha} + Bi \cdot \psi_{s,\alpha} \geq Bi \cdot \psi_s(\alpha x) \tag{B-17}$$

and:

$$2)\ \alpha\psi'_{s,\alpha} + Bi \cdot \psi_{s,\alpha} \geq 0 \tag{B-18}$$

The first inequality can also be written as:

$$Bi \cdot \left[\psi_s(\alpha x) - \psi_{s,\alpha}\right] \geq \alpha\psi'_{s,\alpha} \tag{B-19}$$

This is always true, because the left-hand side of this inequality is always positive or null, whereas the right-hand side is always null or negative. In fact, given that $x_0 \leq x \leq 1$, and therefore $\alpha \cdot x \leq \alpha$, and as, by construction, the function $\psi_s(\alpha x)$ has its maximum value at the stone-to-pulp interface ($\psi_s(\alpha x_0) = 1$ and decreases monotonically to the value at the external surface), $\psi_s(\alpha x) \geq \psi_{s,\alpha}$. Therefore the left-hand side of Eq. (B-19) is always positive or null. And within this range $\psi'_{s,\alpha} \leq 0$, and therefore the right-hand side is always negative or null.

The second inequality indicates that there is a lower threshold Biot number for each value of α. In fact, inequality (B-16) implies that the Biot number must satisfy the condition:

$$Bi \geq -\frac{\alpha \cdot \psi'_{s,\alpha}}{\psi_s(\alpha)} \qquad \text{(B-20)}$$

At the same time, the transcendental equation for the general case is:

$$\delta \psi'_{\delta} = -Bi \cdot \psi_{\delta} \qquad \text{(B-21)}$$

Consequently:

$$Bi = -\frac{\delta \cdot \psi'_{\delta}}{\psi_{\delta}}$$

In other words:

$$-\frac{\delta \cdot \psi'_{\delta}}{\psi_{\delta}} \geq -\frac{\alpha \cdot \psi'_{s,\alpha}}{\psi_s(\alpha)}$$

because the general function:

$$y = \frac{-x\psi'(x)}{\psi(x)}$$

is continuous and monotonically positive within the range $0 \leq x \leq \delta_M$, this condition is only fulfilled when:

$$\alpha \leq \delta$$

This, for convenience, will be expressed in the form:

$$\alpha^2 \leq \delta^2 \tag{B-22}$$

And obviously δ will always be less than its maximum value δ_M (for an infinite Biot number):

$$\delta^2 \leq {\delta_M}^2$$

Consequently, the conditions are finally:

$$\alpha^2 \leq \delta^2 \leq {\delta_M}^2 \tag{B-23}$$

Thus, as the Biot number decreases, α approaches δ and the denominator $\alpha\psi'_{s,\alpha} + Bi \cdot \psi_{s,\alpha}$ will approach $\delta\psi'_{\delta} + Bi \cdot \psi_{\delta}$, which is zero, and therefore in the limit case $\alpha\psi'_{s,\alpha} + Bi \cdot \psi_{s,\alpha} = 0$, and θ_s – and therefore also the complete solution for θ (Eq. (7)) – will tend towards infinity. Following both Jakob's[44] and Carslaw and Jaeger's [15] reasoning, all this means that in such a hypothetical case, with that value of the surface heat transfer coefficient, extraction of heat via the surface would not be sufficient to eliminate the heat generated in the interior and the temperature rise would be unbounded.

Mass Average

The steady state of the mass average in the spherical shell of the pulp is:

$$\bar{\theta}_s = \frac{1}{V}\int_{r=r_0}^{R} \theta_s(x)\, dV \tag{B-24}$$

V being the volume of the spherical shell of the pulp:

$$V = \frac{4}{3}\pi(R^3 - {r_0}^3) \tag{B-25}$$

$$dV = 4\pi r^2 dr \tag{B-26}$$

Therefore, in dimensionless form:

$$\bar{\theta}_s = \frac{3}{(1-{x_0}^3)}\int_{x_0}^{1} x^2\theta_s(x)dx \tag{B-27}$$

As (Eq. (B-8)):

$$\theta_s(\alpha x) = P \cdot \psi_s(\alpha x) - \frac{\beta}{\alpha^2}$$

$$\frac{3}{(1-{x_0}^3)}\int_{x_0}^{1} x^2\theta_s dx = \frac{3 \cdot P}{(1-{x_0}^3)}\int_{x_0}^{1} x^2\psi_s(\alpha x)dx - \frac{\beta}{\alpha^2} \tag{B-28}$$

Taking into account Eqs. (B-6), (B-7):

$$\int_{x_0}^{1} x^2\psi_s(\alpha x)dx = -\frac{1}{\alpha^2}\left[\frac{\partial\psi_s(\alpha)}{\partial x}\right]_{x=1} = -\frac{\psi'_{s,\alpha}}{\alpha} \tag{B-29}$$

Accordingly:

$$\bar{\theta}_s = \frac{3 \cdot P}{(1-{x_0}^3)}\int_{x_0}^{1} x^2\psi_s(\alpha x)dx - \frac{\beta}{\alpha^2} = -\frac{3 \cdot P}{(1-{x_0}^3)}\frac{\psi'_{s,\alpha}}{\alpha} - \frac{\beta}{\alpha^2}$$

Finally, replacing P with its value (Eq. (B-9)):

$$\bar{\theta}_s = -\frac{\beta}{\alpha^2}\left[\frac{3}{\alpha \cdot (1-{x_0}^3)} \cdot \frac{Bi \cdot \psi'_{s,\alpha}}{\alpha\psi'_{s,\alpha} + Bi \cdot \psi_{s,\alpha}} + 1\right] \tag{B-30}$$

When $Bi \to \infty$:

$$\bar{\theta}_{s,\infty} = -\frac{\beta}{\alpha^2}\left[\frac{3}{\alpha\cdot(1-{x_0}^3)}\cdot\frac{\psi'_{s,\alpha}}{\psi_{s,\alpha}} + 1\right] \quad \text{(B-31)}$$

Appendix C. Expansion Constants

General

As seen in Appendix A, the Fourier series solution to Eq. (1) is:

$$Y(x, Fo) = \theta(x, Fo) - \theta_s(x) = \sum_{i=1}^{\infty} J_i \cdot \psi(\delta_i x) \cdot e^{-({\delta_i}^2-\alpha^2)Fo} \quad \text{(C-1)}$$

where:

$$\psi(\delta x) = \frac{\sin[\delta(x-x_0)]}{\delta\cdot x} + \frac{x_0\cdot\cos[\delta(x-x_0)]}{x} \quad \text{(C-2)}$$

is the solution to the equation:

$$\frac{\partial}{\partial x}\left(x^2\frac{\partial\psi(x)}{\partial x}\right) = -\delta^2 x^2 \psi(x) \quad \text{(C-3)}$$

and with:

$$\theta_s(\alpha x) = P \cdot \psi_s(\alpha x) - \frac{\beta}{\alpha^2} \quad \text{(C-4)}$$

$$\psi_s(\alpha x) = \frac{\sin[\alpha(x-x_0)]}{\alpha\cdot x} + \frac{x_0\cdot\cos[\alpha(x-x_0)]}{x} \quad \text{(C-5)}$$

and:

$$P = \frac{Bi\cdot\beta}{\alpha^2\left(\alpha\psi'_{s,\alpha}+Bi\cdot\psi_{s,\alpha}\right)} \quad \text{(C-6)}$$

By construction, at the initial time $= 0\theta(x, 0) = 1$, Eq. (C-1) becomes:

$$\theta(x,0) - \theta_s(x) = 1 - \theta_s = \sum_{i=1}^{\infty} J_i \cdot \psi(\delta_i x) \qquad \text{(C-7)}$$

To calculate J_i, both sides of Eq. (C-7) are multiplied by $x^2\psi(\delta_j x)$ and integrated between $x = 0$ and $x = 1$. This gives:

$$\int_{x_0}^{1} (1-\theta_s) x^2 \psi(\delta_j x) dx = J_i \sum_{i=1}^{\infty} \int_{x_0}^{1} x^2 \psi(\delta_i x) \psi(\delta_j x) dx \qquad \text{(C-8)}$$

For two solutions δ_i and δ_j corresponding to the same Biot number, the following two integral orthogonality properties are deduced:

$$\text{If } \delta_i \neq \delta_j \quad \int_{x_0}^{1} x^2 \psi(\delta_i x) \psi(\delta_j x) dx = 0 \qquad \text{(C-9)}$$

$$\text{If } \delta_i = \delta_j \quad \int_{x_0}^{1} x^2 \psi(\delta_i x) \psi(\delta_j x) dx = \int_{x_0}^{1} x^2 [\psi(\delta_i x)]^2 dx \qquad \text{(C-10)}$$

and, for any α that is not an eigenvalue of the boundary equation (Eq. (A-22)):

$$\int_{x_0}^{1} x^2 \psi_s(\alpha x) \psi(\delta_i x) dx = -\frac{1}{P} \cdot \frac{Bi \cdot \psi(\delta)}{(\alpha^2 - \delta^2)} \cdot \frac{\beta}{\alpha^2} \qquad \text{(C-11)}$$

Therefore:

$$\int_{x_0}^{1} (1-\theta_s) x^2 \psi(\delta_i x) dx = J_i \sum_{i=1}^{\infty} \int_{x_0}^{1} x^2 [\psi(\delta_i x)]^2 dx \qquad \text{(C-12)}$$

and resolving J_i:

$$J_i = \frac{\int_{x_0}^{1} (1-\theta_s) x^2 \psi(\delta_i x) dx}{\int_{x_0}^{1} x^2 [\psi(\delta_i x)]^2 dx} = \frac{I}{\int_{x_0}^{1} x^2 [\psi(\delta_i x)]^2 dx} \qquad \text{(C-13)}$$

Where I is:

$$I = \int_{x_0}^{1} (1 - \theta_s) x^2 \psi(\delta_i x) dx \quad \text{(C-14)}$$

It is easy to verify that:

$$\int_{x_0}^{1} x^2 [\psi(\delta_i x)]^2 dx = \frac{[\psi(\delta_i)]^2 ({\delta_i}^2 + Bi^2 - Bi) - {\delta_i}^2 {x_0}^3}{2{\delta_i}^2} \quad \text{(C-15)}$$

And moreover, introducing Eq. (C-4) into Eq. (C-14):

$$I = \int_{x_0}^{1} (1 - \theta_s) x^2 \psi(\delta_i x) dx = \int_{x_0}^{1} \left(1 + \frac{\beta}{\alpha^2} - P \cdot \psi_s(\alpha x)\right) x^2 \psi(\delta_i x) dx$$

or:

$$I = \left(1 + \frac{\beta}{\alpha^2}\right) \int_{x_0}^{1} x^2 \psi(\delta_i x) dx - P \int_{x_0}^{1} x^2 \psi_s(\alpha x) \psi(\delta_i x) dx \quad \text{(C-16)}$$

As the derivative $[\partial\psi(\delta_i x)/\partial x]$ equals zero at $x = x_0$, the integration in the first addend is:

$$\int_{x_0}^{1} x^2 \psi(\delta_i x) dx = \frac{Bi \cdot \psi(\delta_i)}{{\delta_i}^2} \quad \text{(C-17)}$$

In this way, introducing Eq. (C-11) and Eq. (C-17) in Eq. (C-16) and rearranging terms:

$$I = \frac{Bi \cdot \psi(\delta_i)}{{\delta_i}^2} \left(1 + \frac{\beta}{\alpha^2 - {\delta_i}^2}\right) \quad \text{(C-18)}$$

And finally, introducing Eq. (C-15) and Eq. (C-18) into Eq. (C-13), J_i is:

$$J_i = \frac{2Bi\psi(\delta_i)}{[\psi(\delta_i)]^2({\delta_i}^2+Bi^2-Bi)-{\delta_i}^2{x_0}^3} \cdot \left(1+\frac{\beta}{\alpha^2-{\delta_i}^2}\right) \quad \text{(C-19)}$$

As in the case of the solid with internal linear heat generation [11], constants J_i may be written as:

$$J_i = J_{i,0}\left(1+\frac{\beta}{\alpha^2-{\delta_i}^2}\right) \quad \text{(C-20)}$$

$J_{i,0}$ being the J_i coefficient for the case without internal heat sources [17]:

$$J_{i,0} = \frac{2Bi\psi(\delta_i)}{[\psi(\delta_i)]^2({\delta_i}^2+Bi^2-Bi)-{\delta_i}^2{x_0}^3} \quad \text{(C-21)}$$

When $Bi \to \infty$

$$J_{i,\infty} = J_{i,0,\infty}\left(1+\frac{\beta}{\alpha^2-{\delta_i}^2}\right) \quad \text{(C-22)}$$

$$J_{i,0,\infty} = \lim_{Bi\to\infty}\left[\frac{2Bi\psi(\delta_{i,M})}{[\psi(\delta_{i,M})]^2({\delta_{i,M}}^2+Bi^2-Bi)-{\delta_{i,M}}^2{x_0}^3}\right] \quad \text{(C-23)}$$

Taking into account Eq. (A-34) and rearranging terms yields:

$$J_{i,0,\infty} = \frac{2\sqrt{1+{\delta_{i,M}}^2{x_0}^2}}{1+{\delta_{i,M}}^2{x_0}^2-{\delta_{i,M}}^2{x_0}^3} \quad \text{(C-24)}$$

Accordingly, introducing this in Eq. (C-22):

$$J_{i,\infty} = (-1)^{(i-1)}\frac{2\sqrt{1+{x_0}^2{\delta_{i,M}}^2}}{1+{x_0}^2{\delta_{i,M}}^2-{\delta_{i,M}}^2{x_0}^3}\left(1+\frac{\beta}{\alpha^2-{\delta_{i,M}}^2}\right) \quad \text{(C-25)}$$

Mass Average

As the density is considered to be constant, the mass average of the temperature (from Eq. (C-1)) is:

$$\theta(x, Fo) = \frac{1}{V}\int_{x=x_0}^{1}\left(\theta_s(x) + \sum_{i=1}^{\infty} J_i \cdot \psi(\delta_i x) \cdot e^{-(\delta_i^2-\alpha^2)Fo}\right) dV \quad \text{(C-26)}$$

V being the volume of the spherical shell of the pulp:

$$V = \frac{4}{3}\pi(R^3 - r_0{}^3) \quad \text{(C-27)}$$

$$dV = 4\pi r^2 dr \quad \text{(C-28)}$$

Thus, Eq. (C-26) can be written in dimensionless form, leaving:

$$\bar{\theta}(Fo) = \frac{3}{(1-x_0{}^3)}\int_{x_0}^{1} x^2\theta_s(x)dx + \sum_{i=1}^{\infty} J_i \cdot e^{-(\delta_i^2-\alpha^2)Fo} \cdot \left(\frac{3}{(1-x_0{}^3)}\int_{x_0}^{1} \psi(\delta_i x) \cdot x^2 dx\right) \quad \text{(C-29)}$$

The steady-state mass average temperature $\bar{\theta}_s$ is calculated in Appendix B.

The mass average constants may be written:

$$\bar{J}_i = J_i\bar{\psi}(\delta_i) = J_i\frac{3}{(1-x_0{}^3)}\int_{x_0}^{1} x^2\psi(\delta_i x)dx \quad \text{(C-30)}$$

Introducing Eq. (C-17) in Eq. (C-30):

$$\bar{J}_i = J_i\bar{\psi}(\delta_i) = J_i\frac{3}{(1-x_0{}^3)}\frac{Bi \cdot \psi(\delta_i)}{\delta_i^2} \quad \text{(C-31)}$$

and introducing Eq. (C-19):

$$\bar{J_i} = J_i\bar{\psi}(\delta_i) = \frac{6Bi^2}{\delta^2(1-x_0{}^3)} \cdot \frac{\psi^2(\delta_i)}{[\psi(\delta_i)]^2(\delta_i{}^2+Bi^2-Bi)-\delta_i{}^2x_0{}^3} \cdot \left(1+\frac{\beta}{\alpha^2-\delta_i{}^2}\right)$$ (C-32)

or:

$$\bar{J_i} = \bar{J}_{i,0} \cdot \left(1+\frac{\beta}{\alpha^2-\delta_i{}^2}\right) \quad \text{(C-33)}$$

$$\bar{J}_{i,0} = \frac{6Bi^2}{\delta^2(1-x_0{}^3)} \cdot \frac{\psi^2(\delta_i)}{[\psi(\delta_i)]^2(\delta_i{}^2+Bi^2-Bi)-\delta_i{}^2x_0{}^3} \quad \text{(C-34)}$$

When $Bi \to \infty$

$$\bar{J}_{i,\infty} = \bar{J}_{i,0,\infty} \cdot \left(1+\frac{\beta}{\alpha^2-\delta_{i,M}{}^2}\right) \quad \text{(C-35)}$$

$$\bar{J}_{i,0,\infty} = \lim_{Bi\to\infty}\left[\frac{6}{\delta_{i,M}{}^2(1-x_0{}^3)} \cdot \frac{\psi^2(\delta_{i,M})\cdot Bi^2}{\psi^2(\delta_{i,M})\cdot(\delta_{i,M}{}^2+Bi^2-Bi)-\delta_{i,M}{}^2x_0{}^3}\right]$$ (C-36)

That is:

$$\bar{J}_{i,0,\infty} = \frac{6}{\delta_{i,M}(1-x_0{}^3)} \cdot \frac{1+x_0{}^2\delta_{i,M}{}^2}{1+x_0{}^2\delta_{i,M}{}^2-\delta_{i,M}{}^2x_0{}^3} \quad \text{(C-37)}$$

Appendix D. First Approximations and Applications

From now on, and to facilitate writing, when working exclusively with the first term of the series, and as long as it does not lead to errors, we drop subscript 1, which denotes the order number, and the function's argument whenever this leaves no room for doubts. Therefore:

$$\delta_1 = \delta, J_1 = J \text{ and } \psi_1 = \psi$$

As said in the Introduction, as in the heat conduction in homogeneous solid sphere, in conduction in stone fruits [17] or in conduction with an internal heat source linearly dependent on temperature [11], for practical purposes and from a given moment onwards, the exact solution in infinite series reduces very quickly to its first exponential term, whose representation in semi logarithmic scale is a straight line. The first term of the complete series (Eq. 7) is:

$$\ln Y \approx \ln[J \cdot \psi(\delta \cdot x)] - (\delta^2 - \alpha^2)Fo \qquad \text{(D-1)}$$

$$\ln Y \approx \ln[J] - (\delta^2 - \alpha^2)Fo \qquad \text{(D-2)}$$

This allows us to obtain approximation models for direct calculations (cooling times, for example) or algorithms to indirectly determine thermophysical parameters, constants of the heat of respiration or the surface heat transfer coefficient. Some of these applications or calculation algorithms are shown in this Appendix D.

Direct Question: Prediction of the Chilling Time at the Thermal Center.

Condition: All thermophysical parameters, surface heat transfer coefficient and heat generation constants should be known.

Clearing in Eq. (D-2) the dimensionless time to reach a certain temperature at the pulp-to-stone interface is:

$$Fo(x_0) \approx \frac{\ln J - \ln Y}{(\delta^2 - \alpha^2)} \qquad \text{(D-3)}$$

Thus, the algorithm to be followed is:

1. As the thermophysical parameters, the surface heat transfer coefficient and the external radius are known, the Biot number can be determined.
2. As the heat of respiration constants A_0 and A_1, the thermal conductivity, the external radius and the initial and medium

temperatures are known, the heat source dimensionless numbers (Eqs. (4-6)) can be determined.

3. With these values, and taking into account that at the pulp-to-stone interface $\psi(\delta \cdot x_0) = 1$, the steady state temperature can be determined at the thermal center (Eqs. (9) to (11)):
4. From the Biot number just determined and from the internal radius x_0 (dimensionless radius of the stone) the first eigenvalue of Eq. (12) $\delta = \delta_1$ can be determined.
5. From the Biot number and the heat source dimensionless numbers the numerical series first constant J can be determined (Eq. (13)).
6. Finally, substituting into Eq. (D-3), the corresponding dimensionless time is obtained.

*Displacement Corrections. Simultaneous Determination of the Surface Heat Transfer Coefficient (**h**) and the Slope of the Heat Source (**A**$_1$)*

At the dimensionless distance $x = r/R$ clearing Fo in Eq. (D-1):

$$Fo(x) \approx \frac{\ln(J\psi) - \ln Y}{(\delta^2 - \alpha^2)} = \frac{\ln J + \ln \psi(\delta \cdot x) - \ln Y}{(\delta^2 - \alpha^2)} \tag{D-4}$$

Correction on Time to Get the Same Temperature

To reach the same temperature at x_0 and at the coordinate x, $Fo(x)$ can be rewritten as follows:

$$Fo(x) \approx \frac{\ln J - \ln Y}{(\delta^2 - \alpha^2)} + \frac{\ln \psi(\delta \cdot x)}{(\delta^2 - \alpha^2)} = Fo_c - \frac{\ln(1/\psi(\delta \cdot x))}{(\delta^2 - \alpha^2)}$$

That is, the correction on time for the coordinate x is:

$$D_x \equiv Fo_c - Fo_x \approx \frac{\ln(1/\psi(\delta \cdot x))}{(\delta^2 - \alpha^2)} \tag{D-5}$$

Clearly Eq. (D-5) is dependent not on time but solely on the coordinate. Hence, other than for low Fourier numbers, if we wish to calculate the time required at a coordinate x to reach a given temperature

T, it will suffice to know the time needed to calculate it at the core and adjust this with the appropriate term for displacement.

If $x = 1$ (surface):

$$D_{Sf} \equiv Fo_c - Fo_1 \approx \frac{\ln(1/\psi(\delta))}{(\delta^2 - \alpha^2)} \qquad \text{(D-6)}$$

Correction on Temperature for the Same Time

Here we will distinguish with the subscripts c and x respectively the values corresponding to x_0 and x.

At x_0:

$$Y_c \approx J \cdot e^{-(\delta^2 - \alpha^2)Fo} \qquad \text{(D-7)}$$

At coordinate x:

$$Y_x \approx J \cdot \psi(\delta \cdot x) \cdot e^{-(\delta^2 - \alpha^2)Fo} = \psi(\delta \cdot x) \cdot Y_c \qquad \text{(D-8)}$$

That is, the correction on temperature for the coordinate x is the factor:

$$\Phi_x \equiv \frac{Y_x}{Y_c} \approx \psi(\delta \cdot x) \qquad \text{(D-9)}$$

If $x = 1$ (surface):

$$\Phi_{Sf} \equiv \frac{Y_1}{Y_c} \approx \psi(\delta) \qquad \text{(D-10)}$$

If the two temperature histories are measured, Eqs. (D-5) and (D-9) allows us to the simultaneous determination of $\delta^2$and α^2. This is the ratio method extended to the simultaneous determination of the Biot number and the α^2 number [47].

Consequently, if the thermal conductivity is known, from Eq. (D-5) and Eq. (D-9) the surface heat transfer coefficient h and the slope of the heat source A_1 can be indirectly determined.

Approximation to Maximum Value at the Pulp-to-Stone Interface

As we can see in Figures 2–4, depending on the particular conditions in each case, the temperature may rise at the outset up to a given maximum value Θ_M in a time Fo_M, thereafter falling to its steady value. A first approximation to this value can be achieved by considering that its time derivative must also vanish. Hence, at this point the following condition is required:

$$\left.\frac{dY}{dFo}\right|_{Fo_M} = -\sum_{i=1}^{\infty}\left({\delta_i}^2 - \alpha^2\right)\cdot J_i \cdot \psi(\delta_i x)\cdot e^{-\left({\delta_i}^2-\alpha^2\right)Fo_M} = 0 \qquad \text{(D-11)}$$

At the stone–pulp interface, $\psi(\delta_i x) = 1$:

$$\left.\frac{dY}{dFo}\right|_{Fo_M} = -\sum_{i=1}^{\infty}\left({\delta_i}^2 - \alpha^2\right)\cdot J_i \cdot e^{-\left({\delta_i}^2-\alpha^2\right)Fo_M} = 0 \qquad \text{(D-12)}$$

To calculate a first approximation to this value the first two terms of the series can be taken as significant:

$$\left.\frac{dY}{dFo}\right|_{Fo_M} \approx -\left({\delta_1}^2 - \alpha^2\right)\cdot J_1 \cdot e^{-\left({\delta_1}^2-\alpha^2\right)Fo_M} - \left({\delta_2}^2 - \alpha^2\right)\cdot J_2 \cdot e^{-\left({\delta_2}^2-\alpha^2\right)Fo_M} = 0 \qquad \text{(D-13)}$$

and from this we can find the value of Fo_M:

$$Fo_M \approx \frac{\ln\left[-\frac{\left({\delta_2}^2-\alpha^2\right)\cdot J_2}{\left({\delta_1}^2-\alpha^2\right)\cdot J_1}\right]}{\left({\delta_2}^2-{\delta_1}^2\right)} \qquad \text{(D-14)}$$

If we substitute this value in the sum of the first two terms of the series, we get an approximation to the maximum value attained.

$$\theta_M \approx \theta_{s,x_0} + J_1 \cdot e^{-\left({\delta_1}^2-\alpha^2\right)Fo_M} + J_2 \cdot e^{-\left({\delta_2}^2-\alpha^2\right)Fo_M} \qquad \text{(D-15)}$$

Other Indirect Determinations

General condition: It will be assumed from now on that the experiment is long enough so that the final temperature can be considered as the steady state temperature.

Estimation of the Heat Transfer Coefficient

Condition: All thermophysical parameters and the two heat generation constants are known.

1. Transform the experimental time-temperature history at the thermal center (pulp-stone contact) into a table $\{Fo, Y_i\}$, where $Y_i = (T_i - T_s)/(T_0 - T_{ex})$
2. Obtain the dimensionless number of the heat respiration $\alpha^2 = A_1R^2/k$
3. Obtain the linear regression to the linear portion of its graph in semi-logarithmic coordinates: $\ln Y_c = M - N \cdot Fo \approx \ln[J] - N \cdot Fo$
4. Obtain the value δ^2: $\delta^2 = N + \alpha^2$
5. And solve the Biot number in Eq. (12):

$$Bi = 1 + \frac{\delta^2 x_0 - m}{1 + m x_0} \tag{D-16}$$

Where m is:

$$m \equiv \frac{\delta}{\tan[\delta(1 - x_0)]} \tag{D-17}$$

Finally, clearing in the definition of Biot number, the surface heat transfer coefficient can be calculated:

$$h = \frac{Bi \cdot k}{R} \tag{D-18}$$

Estimation of the Heat Generation Constants

Condition: All thermophysical parameters and the surface heat transfer coefficient are known.

1. As all thermophysical parameters and the surface heat transfer coefficient are known, the Biot number can be calculated ($Bi = hR/k$).
2. From the transcendental boundary Eq. (12) the first eigenvalue δ can be calculated.
3. $\psi(\delta_i)$andJ_0 can be calculated from Eqs. (8) and (14)
4. As the steady state temperature T_s is known, the $\{Y - Fo\}$ table can be deduced and plotted, and from its linear portion, the linear regression can be obtained: $\ln Y_c = M - NFo$.

Identifying to Eq. (D-2) J and α^2 can be determined:

$$J = e^M \tag{D-19}$$

$$\alpha^2 = \delta^2 - N \tag{D-20}$$

5. Taking into account Eqs. (13) and (14) and solving β:

$$\beta = N\left(1 - \frac{e^M}{J_0}\right) \tag{D-21}$$

6. Finally, clearing A_0, A_1 and B_0in Eqs. (4), (5) and (6):

$$A_1 = \frac{\alpha^2 \cdot k}{R^2} \tag{D-22}$$

$$B_0 = \frac{\beta \cdot k \cdot \Delta T_0}{R^2} \tag{D-23}$$

and:

$$A_0 = B_0 - A_1 T_{ex} \tag{D-24}$$

Indirect Measurement of Thermal Diffusivity and Surface Heat Transfer Coefficient

Condition: The thermal conductivity and the two heat generation constants A_0 and A_1 are known.

Similarly to Cuesta and Alvarez [17], the thermophysical parameters and the Biot number can be determined both from the slope and the experimental lag factor$J$.

1. Obtain the dimensionless respiratory constants $\alpha^2$and β.

As thermal conductivity k is known, as well as the respiratory constants A_0y A_1, substitute into Eqs. (4), (5) and (6) to obtain:

$\alpha^2 = \frac{A_1 R^2}{k}$= Dimensionless slope of the heat source

$\beta = \frac{B_0 R^2}{k \cdot \Delta T_0}$= Dimensionless number of the heat source

2. Obtain the experimental dimensionless difference $\theta_{s,exp}$ (from the final portion of the time temperature table) and obtain the $\{Y - t\}$ table at the pulp-stone contact.
3. Obtain the linear regression to the linear portion of its graph in semi-logarithmic coordinates:

$$\ln Y \approx M - S \cdot t$$

4. Identifying Eq. (D-1) the lag factor is directly determined (Eq. (D-19)):

$$J = e^M \tag{D-25}$$

and:

$$S = \frac{a(\delta^2 - \alpha^2)}{R^2} \tag{D-26}$$

5. On the other hand be remembered that (Eqs. 13 and 14):

$$J = J_0\left(1 - \frac{\beta}{\delta^2 - \alpha^2}\right) \tag{D-27}$$

$$J_0 = \frac{2Bi\psi(\delta)}{[\psi(\delta)]^2(\delta^2 + Bi^2 - Bi) - \delta^2 {x_0}^3} \tag{D-28}$$

and:

$$\psi(\delta) = \frac{\sin[\delta(1-x_0)] + \delta x_0 \cos[\delta(1-x_0)]}{\delta} \tag{D-29}$$

6. Bearing in mind that $\alpha^2 \& \beta \ll \delta^2$, for the sole purpose of establishing a first approximate value for δ, the heat source will be omitted and the δ value corresponding to $J = e^M$ (Eq. D-25) will be calculated as indicated in Cuesta and Alvarez [17]. This value will be considered as the starting point for the difference $(\delta^2 - \alpha^2)$in the algorithm for calculating the coefficient of surface heat transmission and thermal diffusivity.
7. By introducing this $(\delta^2 - \alpha^2)$ and β into Eq. (D-27), J_0 is determined.

$$J_0 = \frac{J}{\left(1 - \frac{\beta}{\delta^2 - \alpha^2}\right)} = \frac{e^M}{\left(1 - \frac{\beta}{\delta^2 - \alpha^2}\right)} = J_{0,0} \tag{D-30}$$

8. Take a trial value for δ
9. The corresponding Biot number can be deduced using Eqs. (D-16) and (D-17)
10. Introduce this Biot number and the trial δ into Eqs. (D-28) and (D-29) to obtain J_0

11. If $J_0 = J_{0,0}$ (within the deviation considered), then δ, Bi (i.e., the surface heat transfer coefficient) and the thermal diffusivity a (by substituting in Eq. (D-26) have been simultaneously determined.

If $J_0 \neq J_{0,0}$ (within the deviation considered) then, by increasing or decreasing δ, J_0 can be recalculated until the desired accuracy is obtained. In practice, using a simple algorithm to accelerate the convergence, no more than a few cycles of calculation are needed to determine δ, Bi, and a.

12. With this δ calculate again $(\delta^2 - \alpha^2)$ to back to **7**. When this difference repeats (within the deviation considered) the process is over and the surface heat transfer coefficient h and the thermal diffusivity a have been determined.

Acknowledgments

The authors wish to thank the Institute of Food Science, Technology and Nutrition (ICTAN- CSIC), in which this work was carried out. This research did not receive any specific grant from funding agencies in the public, commercial, or not-for-profit sectors.

References

[1] Erdoğdu, F. (2005). Mathematical approaches for use of analytical solutions in experimental determination of heat and mass transfer parameters. *Journal of Food Engineering*, 68, 233–238.

[2] Erdoğdu, F. (2008). A review on simultaneous determination of thermal diffusivity and heat transfer coefficient. *Journal of Food Engineering*, 86, 453–459.

[3] Erdoğdu, F., Linke, M., Praeger, U., Geyer, M., &Schlüter, O. (2014). Experimental determination of thermal conductivity and

thermal diffusivity of whole green (unripe) and yellow (ripe) Cavendish bananas under cooling conditions. *Journal of Food Engineering*, 128, 46–52.

[4] Kondjoyan, A. (2006). A review on surface heat and mass transfer coefficients during air chilling and storage of food products. *International Journal of Refrigeration*, 29, 863-875.

[5] Cleland, A. C. & Earle R. L. (1982a). A simple method for prediction of heating and cooling rates in solids of various shapes. *International Journal of Refrigeration*, 5, 98-106.

[6] Cleland, A. C., Earle, R. L. (1982b). Freezing time prediction for foods – a simplified procedure. *International Journal of Refrigeration*, 5, 134-140.

[7] Cuesta, F. J., Lamúa, M., & Moreno, J. (1990). Graphical calculation of half-cooling times. *International Journal of Refrigeration*, 13, 317-324.

[8] Cuesta, F. J., Lamúa, M., & Alique, R. (2012). A new exact numerical series for the determination of the Biot number: Application for the inverse estimation of the surface heat transfer coefficient in food processing. *International Journal of Heat and Mass Transfer*, 55, 4053–4062.

[9] Cuesta, F. J., & Lamúa, M. (1995). Asymptotic modelling of the transient regime in heat conduction in solids with general geometry. *Journal of Food Engineering*, 24, 295-320.

[10] Cuesta, F. J., & Lamúa, M. (2002). Calculations of the minimum mean cooling/heating time of general-geometry solids. *Journal of Food Engineering*, 53, 259-271.

[11] Cuesta, F. J., & Lamúa, M. (2009). Fourier series solution to the heat conduction equation with an internal heat source linearly dependent on temperature: Application to chilling fruit and vegetables. *Journal of Food Engineering*, 90, 291-299.

[12] Fikiin, A., & Fikiina, I. (1972). Calculation de la durée de refrigération des produits alimentaires et des corps solides. *Proceedings of the XIIIth International Congress of Refrigeration*, pp. 411-414.

[13] Pflug, I. J., Blaisdell, J. L., & Kopelman, J. (1965). Developing temperature-time curves for objects that can be approximated by sphere, infinite plate or infinite cylinder. *ASHRAE Transactions*, 71, 238-248.

[14] Smith, R. E., Nelson, G. L., & Henrickson, R. L. (1967). Analyses on Transient Heat Transfer from Anomalous Shapes. *Transactions ASAE*, 10, 236-244.

[15] Carslaw, H. S. & Jaeger, J. C. (1959). *Conduction of Heat in Solids* (2nd. Ed). Clarendon Press, Oxford, UK, pp. 246-250.

[16] Helal, M. M. (2012). Generalization of the integral transform method to nonlinear heat-conduction problems in multilayered spherical media. *Journal of King Saud University – Science*, 24, 367–377.

[17] Cuesta, F. J., &Alvarez, M. D. (2017). Mathematical modeling for heat conduction in stone fruits. *International Journal of Refrigeration*, 80, 120-129.

[18] Cinquanta, L., Di Matteo, M., & Estia, M. (2002). Physical pre-treatment of plums (Prunusdomestica). Part 2. Effect on the quality characteristics of different prune cultivars. *Food Chemistry*, 79, 233–238.

[19] Di Matteo, M., Cinquanta, L., Galiero, G., & Crescitelli, S. (2000). Effect of a novel physical pretreatment process on the drying kinetics of seedless grapes. *Journal of Food Engineering*, 46, 83-89.

[20] Di Matteo, M., Cinquanta, L., Galiero, G., & Crescitelli, S. (2002). Physical pre-treatment of plums (Prunusdomestica). Part 1. Modelling the kinetics of drying. *Food Chemistry*, 79, 227–232.

[21] Di Matteo, M., Cinquanta, L., Galiero, G., Crescitelli, S. (2003). A mathematical model of mass transfer in spherical geometry: plum (Prunusdomestica) drying. *Journal of Food Engineering*, 58, 183–192.

[22] Ruiz-López, I. I., Cordova, A. V., Rodríguez-Jimenes, G. C., & García-Alvarado, M. A. (2004). Moisture and temperature evolution during food drying: effect of variable properties. *Journal of Food Engineering* 63, 117–124.

[23] Ruiz-López, I. I., García-Alvarado, M. A. (2007). Analytical solution for food-drying kinetics considering shrinkage and variable diffusivity. *Journal of Food Engineering* 79, 208–216.

[24] da Silva, W. P., Precker, J. W., e Silva, D. D. P. S., e Silva, C. M. D. P. S., & Barbosa de Lima, A. G. (2009). Numerical simulation of diffusive processes in solids of revolution via the finite volume method and generalized coordinates. *International Journal of Heat and Mass Transfer* 52, 4976-4985.

[25] da Silva, W. P., e Silva, C. M. D. P. S., e Silva, D. D. P. S., Neves, G. A., & Barbosa de Lima, A. G. (2010). Mass and heat transfer study in solids of revolution via numerical simulations using finite volume method and generalized coordinates for the Cauchy boundary condition. *International Journal of Heat and Mass Transfer*, 53, 1183–1194.

[26] Hernández-Díaz, W. N, Ruiz-López, I. I., Salgado-Cervantes, M. A., Rodríguez-Jimenes, G. C., & García-Alvarado, M. A. (2008). Modeling heat and mass transfer during drying of green coffee beans using prolate spheroidal geometry. *Journal of Food Engineering*, 86, 1–9.

[27] Fabbri, A., Cevoli, C., Alessandrini, L., Romani, S. (2011). Numerical modeling of heat and mass transfer during coffee roasting process. *Journal of Food Engineering*, 105, 264–269.

[28] Nilnont, W., Thepa, S., Janjai, S., Kasayapanand, N., Thamrongmas, C., Bala, B. K. (2012). Finite element simulation for coffee (Coffeaarabica) drying. *Food and Bioproducts Processing*, 90, 341–350.

[29] Ruiz-López, I. I., Ruiz-Espinosa, H., Luna-Guevara, M. L., García-Alvarado, M. A. (2011). Modeling and simulation of heat and mass transfer during drying of solids with hemispherical shell geometry. *Computers & Chemical Engineering*, 35, 191–199.

[30] Ruiz-López, I. I., Ruiz-Espinosa, H., Arellanes-Lozada, P., Bárcenas-Pozos, M. E., García-Alvarado, M. A. (2012). Analytical model for variable moisture diffusivity estimation and drying simulation of

shrinkable food products. *Journal of Food Engineering*, 108, 427–435.

[31] Sabarez, H. T. (2012). Computational modelling of the transport phenomena occurring during convective drying of prunes. *Journal of Food Engineering*, 111, 279–288.

[32] Uyar, R., & Erdoğdu, F. (2012). Numerical evaluation of spherical geometry approximation for heating and cooling of irregular shaped food products. *Journal of Food Science*, 77, E166-E175.

[33] Campañone, L. A., Giner, S. A, & Mascheroni, R. H. (2002). Generalized model for the simulation of food refrigeration. *International Journal of Refrigeration*, 25, 975–984.

[34] Tanner, D. J., Cleland, A. C., Opara, L. U., Robertson, T. R. (2002a). A generalized mathematical modelling methodology for the design of horticultural food packages exposed to refrigerated conditions. Part 1. Formulation. *International Journal of Refrigeration*, 25, 43–53.

[35] Tanner, D. J., Cleland, A. C., & Opara, L. U. (2002b). A generalized mathematical modelling methodology for the design of horticultural food packages exposed to refrigerated conditions. Part 2. Heat transfer modelling and testing. *International Journal of Refrigeration*, 25, 43–53.

[36] Dincer, I. (1994). Unsteady heat-transfer analysis of spherical fruit to air flow. *Energy*, 19, 117–123.

[37] Dincer, I. (1997). *Heat Transfer in Food Cooling Applications* (Taylor and Francis, Washington, Chapter 5) pp. 204-233.

[38] Meffert, H.F. Th., Rudolphij, J.W., & Rooda, J.W. (1971). *Proceedings of the XIII Congress of Refrigeration*, 2, 379-385.

[39] Mendonça, S. L. R., Filho, C. R. B., & da Silva, Z. E. (2005). Transient conduction in spherical fruits: method to estimate the thermal conductivity and volumetric thermal capacity. *Journal of Food Engineering*, 67, 261–266.

[40] Wang, S., Tang, J., Cavalieri, R. P. (2001). Modeling fruit internal heating rates for hot air and hot water treatments. *Postharvest Biology* and *Technology*, 22, 257–270.

[41] SadashiveGowda, B., Narasimham, G. S. V. L., & Krishna Murthy, M. V. (1997). Forced-air precooling of spherical foods in bulk: a parametric study. *International Journal* of *Heat* and *Fluid Flow*, 18, 613–624.

[42] Kole, N. K., &Prasad, S. (1994). Respiration rate and heat of respiration of some fruits under controlled atmosphere conditions. *International Journal of Refrigeration*, 17, 199–204.

[43] Rao, N., Flores Rolando, A., & Gast Karen, L. B. (1993). Mathematical relationships for the heat of respiration as a function of produce temperature. *Postharvest Biology* and *Technology*, 3, 173–180.

[44] Jakob, M. (1949). *Heat Transfer* (John Wiley & Sons, Inc., New Jersey) Chapter 10, Section 10-4.

[45] Gröber, H., Erk, S. & Grigull, U. (1967). *Transmisión del Calor* [*Heat Transmission*], First ed., Selecciones Cientificas, Madrid, pp. 15–17.

[46] ASHRAE. (2010). Thermal properties of foods. ASHRAE® Handbook – Refrigeration (SI), American *Society of Heating, Refrigerating and Air Conditioning Engineers*, Chapter 19, pp. 19.1-19.31.

[47] Awuah, G. B., Ramaswamy, H. S., & Simpson, B. K. (1995). Comparison of two methods for evaluating fluid-to-surface heat transfer coefficients. *Food Research International*, 28, 261-271.

In: Agricultural Research Updates. Volume 27 ISBN: 978-1-53615-916-5
Editors: Prathamesh Gorawala et al.

Chapter 6

HOW PARAMETERS OF WHEAT TECHNOLOGICAL QUALITY AFFECT THE BAKING PROPERTIES IN YEARS WITH DIFFERENT WEATHER CONDITIONS

Dragan R. Živančev*[1,*]*, Jasna S. Mastilović*[2]*,
***Aleksandra M. Torbica*[2]*, Vojislava M. Momčilović*[1]**
***and Milan M. Mirosavljević*[1]**
[1]Institute of Field and Vegetable Crops,
Small Grains Department, Novi Sad, Serbia
[2]University of Novi Sad, Institute of Food Technology
in Novi Sad, Novi Sad, Serbia

ABSTRACT

Extreme weather conditions have been occurring every year in certain regions of the world since the beginning of the 21st century; dry and rainy years alternate each other, and tEere are long periods with

---

[*] Corresponding Author's E-mail: dragan.zivancev@ifvcns.ns.ac.rs.

extremely high temperatures. Agricultural crop production, with its main task to feed the humankind, has definitely been affected by them. The effects of climate changes on crop yields are the general focal point of research studies, while crop quality is much less under the spotlight. This is especially true for wheat due to many parameters of its technological quality, the most important of which are the baking properties. The aim of this study was to analyse the effects of certain parameters of technological quality on the baking properties of wheat in years with different weather conditions. The results from this study showed that there were statistically significant correlation coefficients between protein content, wet gluten content, alveograph properties (work, extensibility and tenacity and extensibility ratio), exstensograph properties (energy, resistance and extensibility ratio) and bread loaf volume in the year with dry and hot weather from anthesis to full grain maturity. The same parameters were not correlated with bread loaf volume in the year with high precipitation from anthesis to full grain maturity.

Keywords: weather conditions, baking properties, technological quality

INTRODUCTION

Extreme weather conditions have been occurring every year in certain regions of the world since the beginning of the 21st century; dry and rainy years alternate each other, and there are long periods during a growing season with extremely high temperatures. Agricultural crop production, with its main task to feed the humankind, has definitely been affected by those unpredictable growing seasons. Drought (Lansigan, 2000), heat stress (Lobell and Gourdji, 2012), rainfall (Patil et al., 2013) and hailstone (Pandey, 2017) have substantial economic impact on food production worldwide. The effects of climate changes on crop yields are currently the general focal point of research studies (Bocchiola et al., 2009; He et al., 2018; Liu et al., 2017; Shrestha et al., 2017; Webber et al., 2015; Wei et al., 2014; Zimmermann et al., 2017), while crop quality is much less under the spotlight.

Climate changes and weather conditions significantly affect the development of wheat plant and kernel, composition of gluten and end use quality, and therefore represent a major issue in different studies. In

Europe average annual air temperature has increased by 0.9°C in the last century (Alcamo et al., 2007). Recently, temperature increase is even more pronounced; 0.4°C for the period from 1977 to 2001 (Jones and Moberg, 2003). Projection of climate change by the end of the 21st century indicates that summer temperature will increase, especially in the south-western parts of Europe (France and the Iberian Peninsula) (Giorgi et al., 2004), and in some regions even more than 6°C. On the other hand, rainfall will vary significantly from year to year on the whole continent (Christensen and Christensen, 2007). These projections for Serbia from the beginning of the 21st century to year 2030 indicate that average daily temperature and average annual precipitation is projected to be from 1.2 to 1.6°C and 10 mm in the northern and eastern parts of Serbia, respectively. Unfortunately, in the central and western parts of the country the amount of the average annual precipitation is projected to decrease by the same amount (Ruml et al., 2012). Also, in the last 30 years of the 21st century, increase of average daily temperature in Serbia is projected to be from 4.0 to 4.4°C and decrease of average precipitation per year from 5 to 50 mm (Ruml et al., 2012).

The most important weather condition parameters affecting wheat crop and kernel development, gluten composition and end use quality are high temperature (Altenbach et al., 2003; Daniel and Triboi, 2000; Debaeke et al., 1996; Don et al., 2005; DuPont et al., 2000; Gibson et al., 1998; Gil et al., 2011; Panozzo and Eagles, 2000; Spiertz et al., 2006; Stone and Savin, 1999; Tashiro and Wardlaw, 1990) and drought (Altebach et al., 2003; Borghi et al., 1997; Carceller and Aussenac, 2001; Daniel and Triboi, 2000; Debaeke et al., 1996; Garrido–Lestache et al., 2004; Gooding et al., 2003). There are also studies on the influence of rainfall on kernel development, gluten composition and end use quality. These studies analysed only a few parameters of breadmaking properties of wheat flour and kernel (Basso et al., 2012; Garrido–Lestache et al., 2004; López–Bellido et al., 2001). The impact of increased CO_2 content in the atmosphere (Högy and Fangmeier, 2008) and the reduction of ozone layer (Picchi et al., 2010) on plant development and properties of technological quality of wheat have been least studied.

As mentioned above, weather conditions and climate change affect breadmaking properties of wheat. Increase in total amount of precipitation is positively correlated with test weight (Garrido–Lestache et al., 2004). Drought in the grain filling stage in Italy in 1992 led to the formation of shrivelled grains, which caused a decrease in test weight for 11 kg/hl (Borghi et al., 1997). Schuler et al., (1995) proved that even though shrivelled grains were removed from the wheat before milling, weather conditions still affected the technological quality. Pomeranz et al., (1985) found that grain hardness is changeable depending on the weather conditions. Temperature, insolation and duration of grain filling affect protein content of grain. According to Johnson and Mattern (1987), these weather parameters in northwestern Europe extended the duration of the grain filling stage. The result was negative association between grain weight and protein content. Contrary to this, Zhao et al., (2009) found that the impact of artificially induced drought did not increase protein content in wheat grain. In areas with warm and droughty climate, such as the USA highland and the Mediterranean region, protein content in wheat kernel is mostly high. Several studies have shown that nitrogen concentration in wheat grains increases when the temperature increases, and rainfall decreases (Daniel and Triboi 2000; Debaeke et al., 1996). However, a few studies have found that large amounts of precipitation in spring months of April and May (Garrido-Lestache et al., 2004) and May (López-Bellido et al., 2001) increase protein content in wheat. Also, Garrido–Lestache et al., (2004) found that weather conditions significantly affected the gluten index. Huge differences in the meteorological parameters from different years resulted in significant year-by-year variation of gluten index values. Relatively warm and dry weather in 2008 in Israel caused a significant decline in gluten index values (Gil et al., 2011). Gooding et al., (2003) found that drought-induced stress occurring before grain filling stage significantly affected SDS sedimentation, more so than heat stress or drought in the subsequent stages of wheat development. When Garrido–Lestache et al., (2004) monitored the influence of weather conditions in different years, they found that impact of these parameters on rheological properties of wheat dough was not clearly defined as in the case of protein content. Randall and Moss (1990) found that temperatures up to 30°C during the grain filling stage increase dough strength, whereas temperatures above this value decrease dough strength. The temperature limit for durum wheat is 35°C according to Borghi et al., (1995). Ciaffi et al., (1996) indicated that even a short period of very high temperatures (35-40°C) during grain development has a negative impact on the technological quality of wheat. The high average values of the tenacity and extensibility ratio (P/L) alveograph parameter (1.49) obtained by Garrido-Lestache et al., (2004), indicate that high temperature in May and low relative humidity during the grain filling stage accelerate grain ripening. Also, these affect the glutenin and gliadin ratio in wheat endosperm (Borghi et al., 1997). López–Bellido et al., (2001) found that alveograph work (W) did not significantly vary in years with highest amount of rainfall and wheat yield, when increased amount of nitrogen fertilizer was applied.

Heat stress increased the Hagberg falling number to 400 s when wheat was exposed to its effect in the period of 15 to 28 days after anthesis (Gooding, 2003). As for baking properties, heat stress affects bread volume. Peterson et al., (1998) found that increase in temperature up to 32°C positively affected bread volume. If the period with temperatures above 32°C lasts longer than 90 h, the bread volume begins to decrease considerably.

Methods for Examination of Wheat Technological Quality

The properties of wheat dough are measured by indirect parameters, whereas the properties of final products are measured by direct parameters. The wheat dough has always been the focus of studies. Wheat flour, out of all existing plant species, possesses a feature that during mixing with water forms dough of unique physico-chemical properties, and dough characteristics depend on the plastic-elastic properties of gluten. Initially, the empirical rheology of the wheat dough was determined by farinograph, extensograph, amylograph and mixograph; whereas from the 1990s alveograph (Pena et al., 1990) started being used more regularly.

Amylograph is a device for indirect examination of flour amylolytic activity, but the main disadvantage of this device is the duration of the analysis. Amylograph is most used in baking industry for examination of amylolytic activity of refined flour, whereas the Hagberg falling number method has become the standard method for determination of the amylolitic activity of wheat cultivars. However, according to Konopka et al., (2004) it can still be efficiently used for different purposes. Also, Indrani et al., (2011) used this device to examine the influence of grains originated from different crops on properties of traditional Indian type of flat bread. Cato and Mills (2008) used the results gained by amylograph to compare with results of amylolytic activity of wheat flours gained by Mixolab.

Extensograph is used for examination of dough strength, energy, resistance and extensibility (Kaluđerski and Filipović, 1998). The main disadvantages of this method are the duration of the analysis and large amount of samples (usually 300 g of sample is necessary per analysis). The importance of the extensogaph is reflected in the study of Miś and Dziki (2013). They used the appearance and values of parameters from extensograph curve to obtain the mathematical model for predicting the influence of dietary fibers on the properties of wheat dough. Furthermore, Indrani et al., (2011) monitored the negative effect of a mixture of different grain cereals on the dough strength through the decrease of the extensograph resistance to extension, extensibility and energy. Also, Zhang et al., (2007) used extensoraph to investigate the influence of protein fractions of bread wheat on the empirical rheology and the quality of the traditional Northern Chinese steam bread.

Alveograph is a device where the dough is subjected to bi-axial stretching, meaning it grows into two normal directions of the same plane, similarly to the air expanding in the spherical balloon. This is the main difference in comparison to the extensograph where dough is subjected to uniaxial stretching (Dobraszczyk and Morgenstern, 2003). Similar to extensograph, alveograph is used to determine the basic properties of dough strength such as deformity W, dough tenacity (P) and extensibility (L) (Kaluđerski and Filipović, 1998). In the early 20th century, the blowing of air into dough was used as an empirical technique for measuring the extensibility of wheat gluten and dough (Chopin, 1921). This method was improved by development of the alveograph device for examination of rheological properties of dough (Launay et al., 1977) as well as measuring of bursting and biaxial rheological properties of wheat dough and gluten, for determination of end use quality (Dobraszczyk and Roberts, 1994; Dobraszczyk, 1997). The main advantage of this rheological method is that the deformation to which the dough is subjected to is almost identical to the deformation that occurs during the dough fermentation when yeast produces CO_2. In the available studies, alveograph was used to examine the technological quality of various wheat genotypes (Dumur et al., 2010; Ktenioudaki et al., 2011; Pena et al., 1990; Rakszegi et al., 2008).

Gluten index is another method for determination of gluten quality. The method is positively correlated with empirical parameters of dough rheology - extensograph resistance and energy (Ćurić et al., 2001). The method is fully automated, reducing the impact of human error in determining this parameter (Dowell et al., 2008; Gil et al., 2011). At the same time, this method provides the data about the content of wet gluten from flour and the qualitative properties of gluten strength through the gluten index value (amount of gluten residue remains of the sieves after centrifugation expressed percentage).

Breadmaking properties are assessed after the bread baking and the following parameters measured: bread volume, colour of bread crumb and crust, and texture of bread crumb.

Although values of rheological parameters to a certain extent define the quality of flour and dough, these are, however, indirect properties of the end use quality of wheat. After assessment of wheat baking properties, it can be determined whether the raw material is suitable for bread making. The earliest studies have shown that the bread volume of loaves prepared from the flour with certain ash content correlates with the protein content (Aitken and Geddes, 1934). When different amounts of dry gluten are added to flour with low-protein content, statistically significant positive correlations are found between the amount of added dry gluten and the strength of dough and bread loaf volume (Aitken and Geddes, 1939). Also, increase in extraction of flour during wheat milling process negatively affects the bread volume (Pomeranz et al., 1977).

In order to evaluate the comprehensive quality of bread, it is necessary to determine its texture properties. Texture properties of bread crumb are mostly evaluated 24 h after baking. Bread firmness is one of the most important texture parameters. The American Association of Cereal Chemists (AACC) standardized this method by using "Universal testing machine" (AACC 74–09). Furthermore, this method is used when it is necessary to examine influence of gliadin and glutenin on the breadmaking properties of wheat (Barak et al., 2013), or to evaluate the addition of β-glucan to wheat flour on baking quality (Skendi et al., 2010). Ktenioudaki et al., (2011) determined baking quality of Irish wheat cultivars by using of Texture Analyser TA–XT2i (TPA analysis). TPA analysis allows determination of several texture parameters. From all measured parameters, Ktenioudaki et al., (2011) used the bread hardness to calculate the internal hardness of bread slices. Also, this property was used together with the measured values of bread volume and digital image analysis to determine the quality of wheat cultivars.

About Our Research

In this study, wheat cultivars used for the examination were grown in three growing seasons (2008, 2009 and 2010). In May of the first production season, insolation and mean temperature were above average values, whereas the amount of precipitation was below 25-year average (Figure 1) (hot season). However, in May of the third production season, the mentioned parameters showed the opposite trend (wet season). Quite similar situation was observed in June of the first and the third production seasons, except for the amount of precipitation in the first production season where the value was on the 25-year average level (Figure 2).

Weather conditions in May of the second production season were quite similar to the first production season, whereas the weather conditions in June of the second season were on the level similar to 25-year average (mixed season). Small differences in values of weather conditions were recorded in July of all three production seasons (Figure 3).

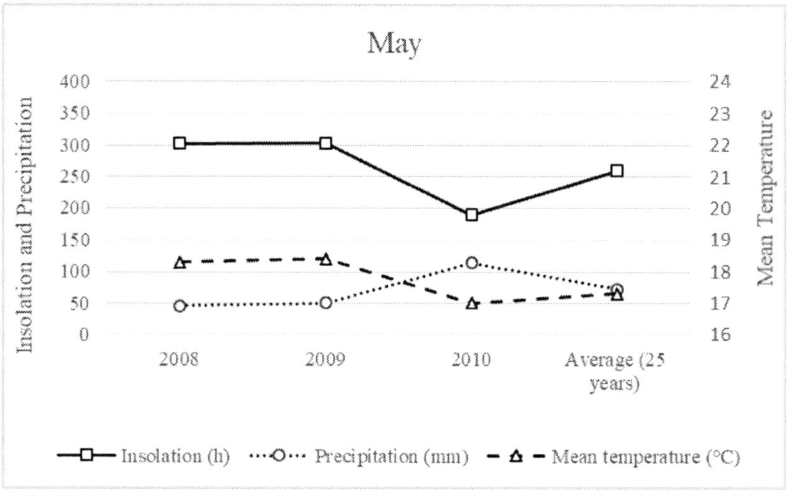

Figure 1. Comparison of weather conditions for May in 2008, 2009 and 2010.

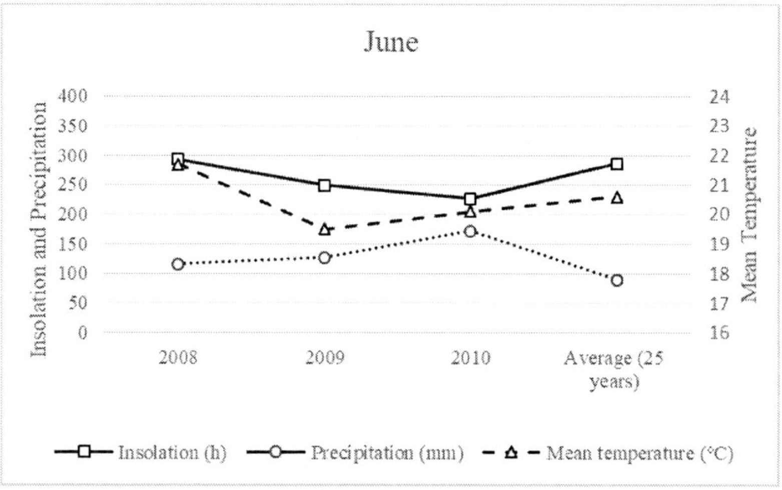

Figure 2. Comparison of weather conditions for June in 2008, 2009 and 2010.

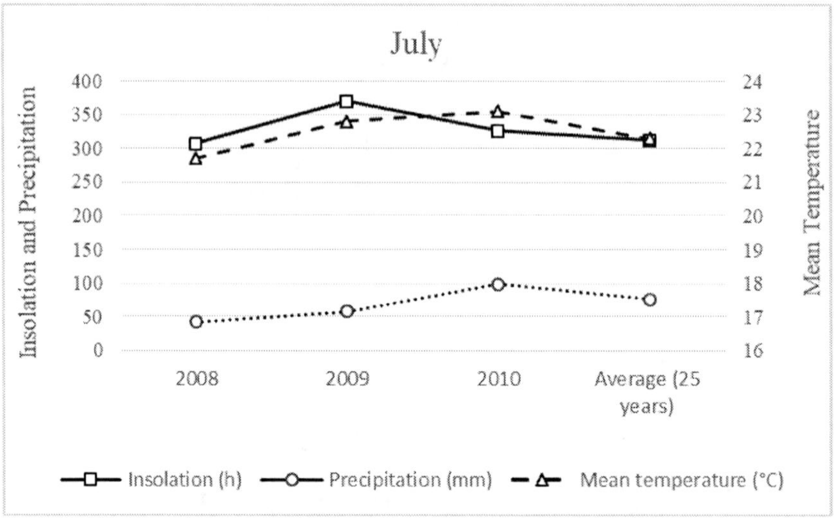

Figure 3. Comparison of weather conditions for July in 2008, 2009 and 2010.

Trials were carried out on 17 bread wheat cultivars produced in all three seasons. Wheat samples were milled using an MLU-202 laboratory mill (Bühler, Uzwil, Switzerland). From the obtained flow stream flours, flour containing 60% starting wheat was formed.

Protein content of wheat grains on dry matter basis (d.m. basis) were determined according to standard analytical (AACC, 2000) methods 52 44-15.02 and 46-16.01. Rheological properties of dough were determined according to, standard methods (ICC, 1996) 114/1 and 121 using an extensograph (CW Brabender, Duisburg, Germany) and an alveograph (Chopin, Paris, France) respectively. Wet gluten content and falling number were determined according to the standard method (106/2 and 107/1, respectively). The content of damaged starch was determined by the amperometric method (SD Matic, Chopin, Paris, France) and calculated using AACC method 76-31.

The breadmaking process was conducted by closely following the procedure described at study Mastilović et al., (2017). After baking, bread loaves were cooled at room temperature for 1 h and stored in controlled conditions at 22 ± 0.7°C and 75 ± 0.5% RH for 23 h before bread quality was assessed. The loaf volume was measured by millet displacement.

Breadcrumb texture profile analysis (TPA) was performed using a Texture Analyser (model TAXT2i, Stable Micro Systems, Godalming, UK) by closely following the procedure described in Mastilović et al., (2018). Slices of bread crumb were scanned (300 dpi) and analysed by ImageJ software. The parameters which were estimated by the software are a number of pores in bread crumb (Valid N) and percentage of total area (PA). It is software package freely downloadable from the US National Institute of Health website (http://imagej.nih.gov/ij/).

Linear regression analysis and correlation matrix (Pearson) analysis were performed by XLSTAT-Pro (demo version, Version 3.02, 2009) software.

Correlations between Technological Quality of Wheat and Different Weather Conditions

One of the most examined correlations regarding wheat technological quality is certainly between grain protein content and bread volume. In various studies (Tipples and Kilborn, 1974; Rubenthaler and Pomeranz, 1987; Ohm et al., 1998) these properties are significantly correlated. In all three seasons the correlations were expectedly positive (Figure 4). Also, the coefficient of determination in the hot season was statistically significant. However, in mixed and wet seasons the estimated regression did not fit well, since coefficients of determination were not significant.

Wet gluten content is a more reliable parameter for prediction of bread volume than protein content in wheat kernel, since wet gluten is a compound composed mostly from glutenins and gliadins. Similarly to protein content, wet gluten also significantly correlates with bread volume (Tipples and Kilborn, 1974; Rubenthaler and Pomeranz, 1987; Ohm et al., 1998). Also, addition of gluten in different forms can increase the bread volume (Czuchajowska and Paszczynska, 1996). In hot and mixed seasons the correlations were positive, whereas in the wet season they were negative (Figure 5). Similarly to protein content, the correlation between bread volume and wet gluten in the hot season was significant, whereas in mixed and wet seasons the estimated regression model did not fit well.

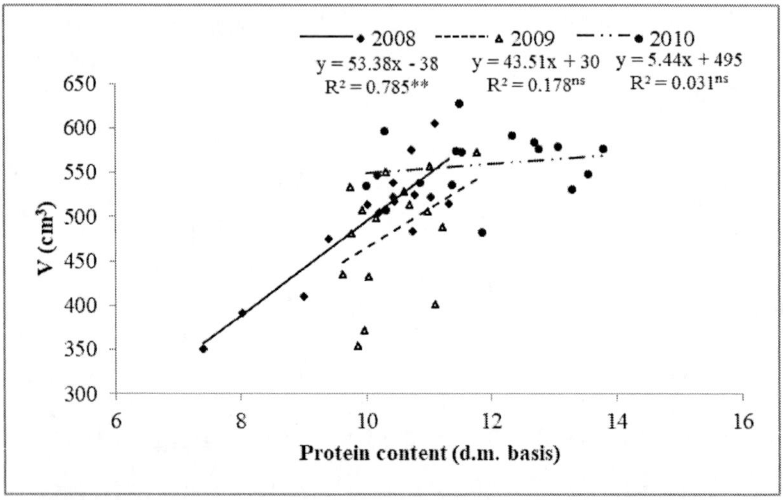

Figure 4. Linear regression between bread volume and protein content in three different production seasons (* significant at 0.01 probability level; ** significant at 0.05 probability level; ns - not significant).

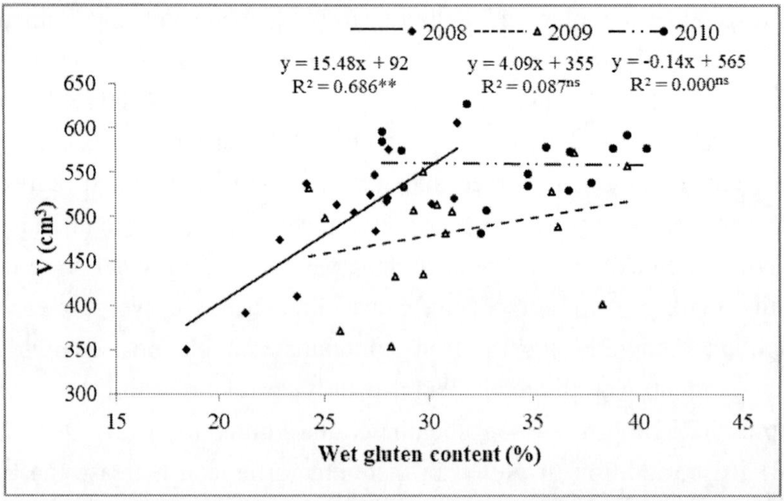

Figure 5. Linear regression relationship between bread volume and wet gluten in three different production seasons (* significant at 0.01 probability level; ** significant at 0.05 probability level; ns - not significant).

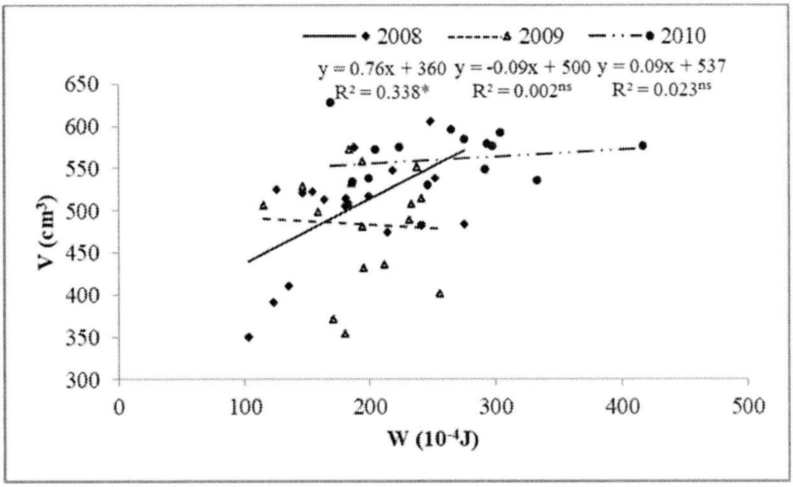

(a)

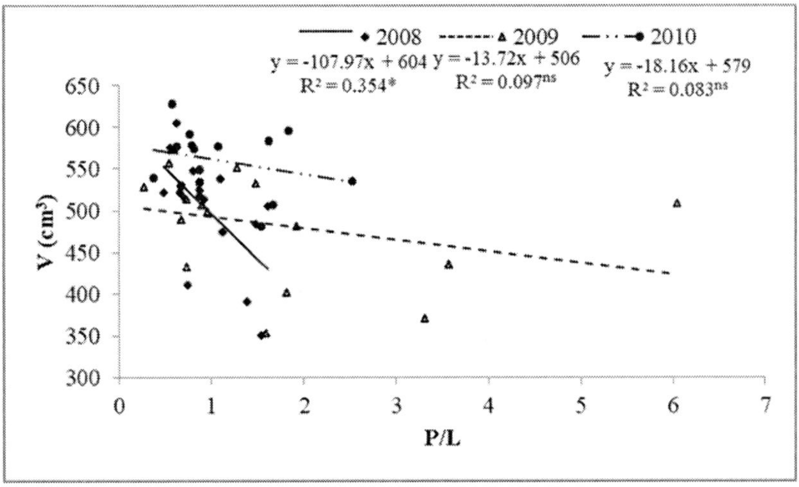

(b)

Figure 6. Linear regression relationship between bread volume and alveograph work (W 10^{-4} J) (A) and bread volume and alveograph (P/L) (B) in three different production seasons (* significant at 0.01 probability level; ** significant at 0.05 probability level; ns - not significant).

As previously mentioned, the parameters of dough strength are very important for determination of breadmaking quality. Various studies (Bralard et al., 1991; Færgestad et al., 2000; Janssen et al., 1996) showed positive correlations between the alveograph work (W 10^{-4} J) and bread volume. In the hot season it was confirmed, since the estimated regression model fit well (Figure 6 A), but it was not the case in mixed and wet seasons because of the non-significant values of coefficients of determination (R^2). The same was observed in the case of correlations between bread volume and alveograph parameter (P/L) (Figure 6 B). Parameter of alveograph extensibility showed good fitting of linear regression model in hot and mixed seasons (Figure 7) whereas correlations were positive in all three seasons.

In the case of uniaxial measurement of dough strength, the correlations between bread volume and extensograph parameters were different. Positive significant correlation between extensograph extensibility and specific bread volume was found only in the hot season (Figure 8), similarly to Příhoda (1980) in Chech harvests of 1978 and 1979. Although relationships were positive in all three seasons, the coefficients of determination in mixed and wet seasons were not significant. The same situation was observed with relationship between bread volume and resistance and extensibility ratio (Figure 9).

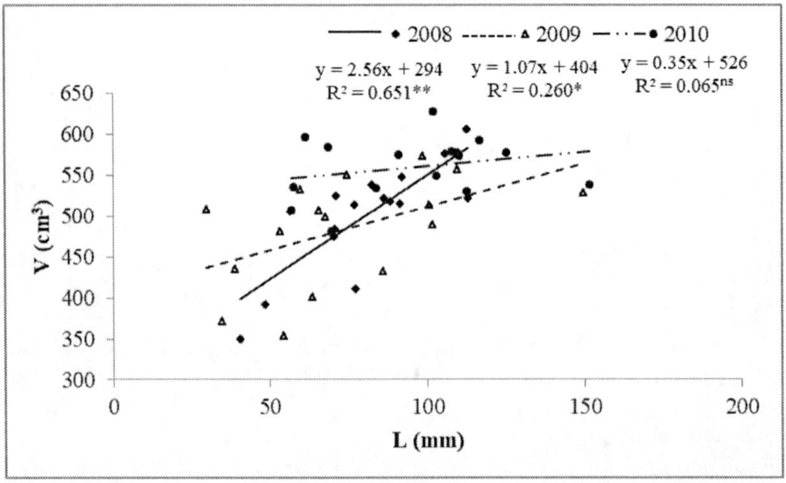

Figure 7. Linear regression relationship between bread volume and alveograph extensibility (L) in three different production seasons (* significant at 0.01 probability level; ** significant at 0.05 probability level; ns - not significant).

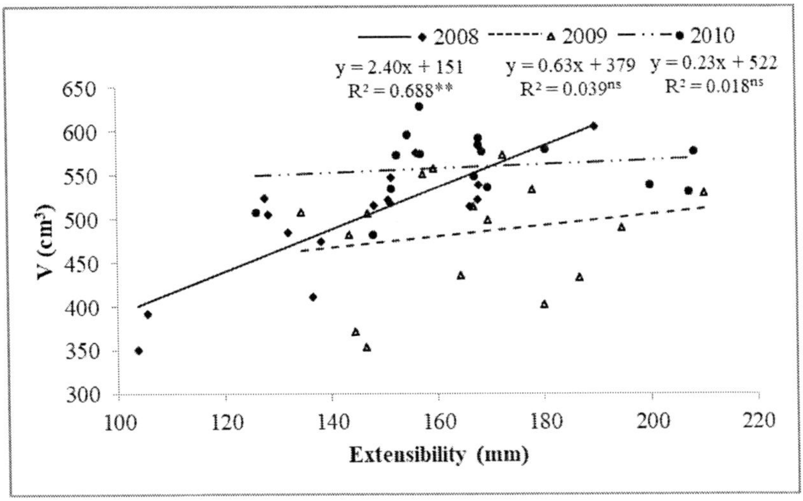

Figure 8. Linear regression relationship between bread volume and extensograph extensibility in three different production seasons (* significant at 0.01 probability level; ** significant at 0.05 probability level; ns - not significant).

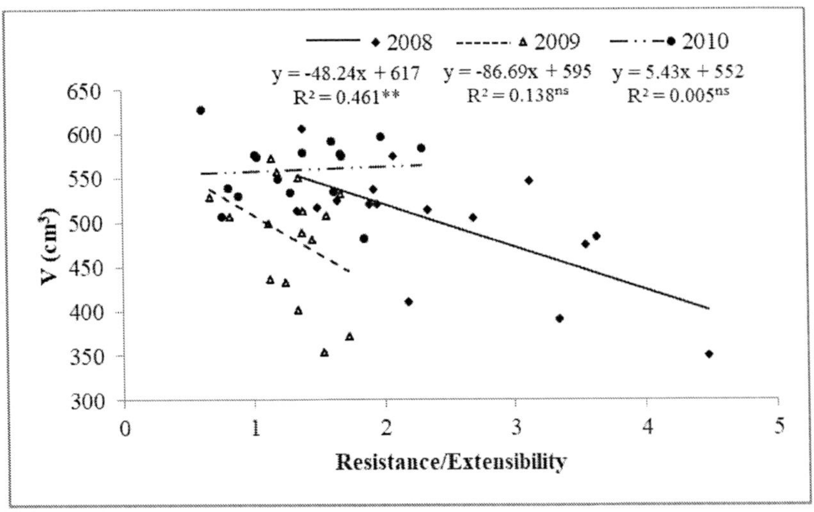

Figure 9. Linear regression relationship between bread volume and extensograph resistance and extensibility ratio in three different production seasons (* significant at 0.01 probability level; ** significant at 0.05 probability level; ns - not significant).

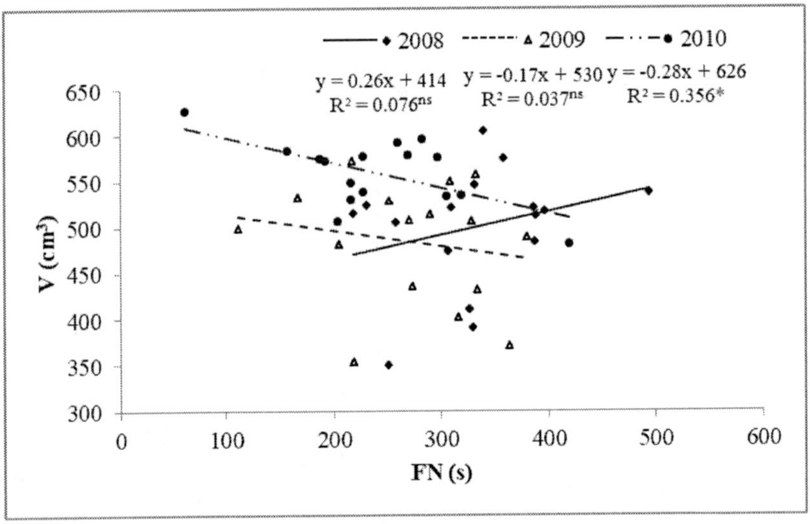

Figure 10. Linear regression relationship between bread volume and FN in three different production seasons (* significant at 0.01 probability level; ** significant at 0.05 probability level; ns - not significant).

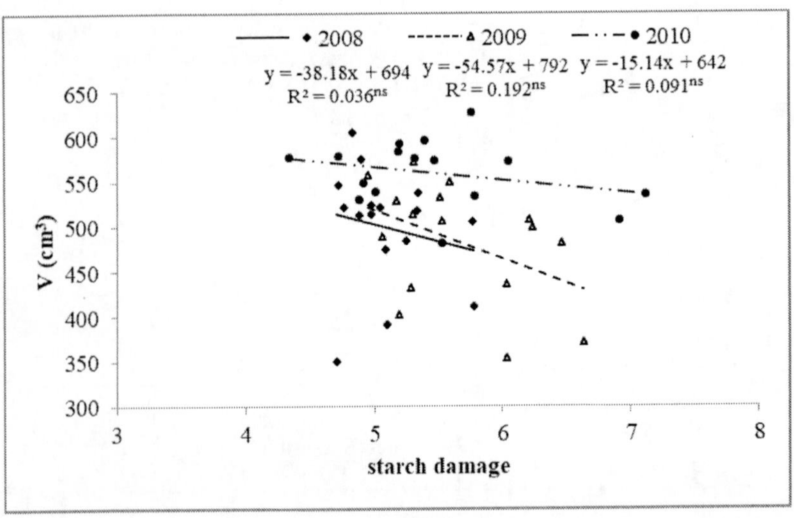

Figure 11. Linear regression relationship between bread volume and starch damage in three different production seasons (* significant at 0.01 probability level; ** significant at 0.05 probability level; ns - not significant).

According to Czubaszek et al., (2001) and Różyło and Laskowski (2011), falling number (FN) was shown to be a good predictor of breadmaking properties. In the hot season, these two parameters were positively correlated, but in the mixed season they were negatively correlated (Figure 10). The linear regression models did not fit well in these two seasons, since coefficients of determination (R^2) were not statistically significant. Unexpectedly, in the wet season the linear regression model fit well (significant R^2) and the correlation was positive. Even though the values of FN were below the optimal 250, the values of bread volume were high (Figure 10).

According to Preston et al., (1987) increase in starch damage in hard red soft wheat flours effects on parameter of dough strength measured by alveograph. Namely, L of dough decreased and P increased. Considering that according to Addo et al., (1990) coefficient of correlation between L and bread volume was statistically positive significant and between P and bread volume was not, the increase in starch damage should lead to decrease in the bread volume. The linear models for all three production season (Figure 11) confirmed this trend, but coefficients of determination (R^2) were not statistically significant.

Bread crumb hardness showed negative relationship with bread volume in all three production seasons (Figure 12). Also, coefficients of determination were all statistically significant. These results are similar to results from study of Al-Saleh and Brennan (2012) who found negative statistically significant coefficient of correlation between bread loaf volume and bread crumb firmness of sliced bread immediately after baking (1 h of slices resting) and 2 days of slices resting. Contrary to bread crumb hardness, cohesiveness showed correlation with bread volume (Figure 13). In hot and mixed seasons coefficients of determinations were statistically significant, whereas in wet season it was not the case.

Similar to the correlation between TPA properties of bread crumb and area of pores (PA (%)), it was found that linear regression models fit well in hot and mixed seasons for bread crumb hardness and bread crumb springiness (Figures 14 and 15). Coefficients of determination were statistically significant. However, correlations in all three years between bread volume and bread crumb hardness were negative, whereas those between bread volume and bread crumb springiness were positive in hot and mixed seasons.

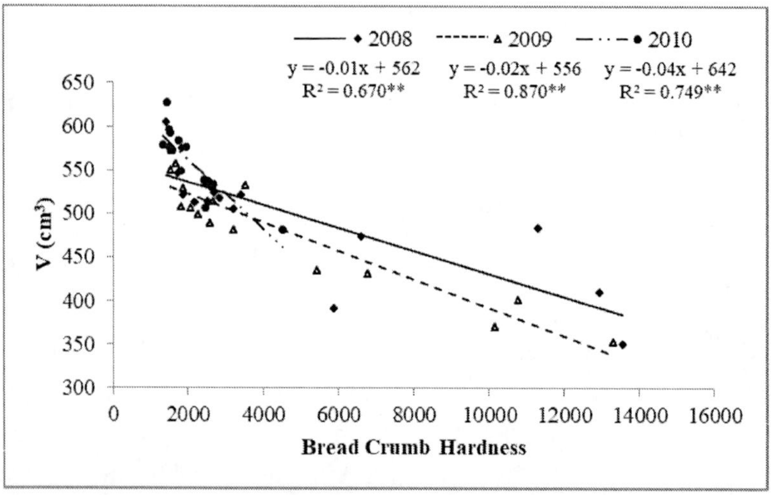

Figure 12. Linear regression relationship between bread volume and bread crumb hardness in three different production seasons (* significant at 0.01 probability level; ** significant at 0.05 probability level; ns - not significant).

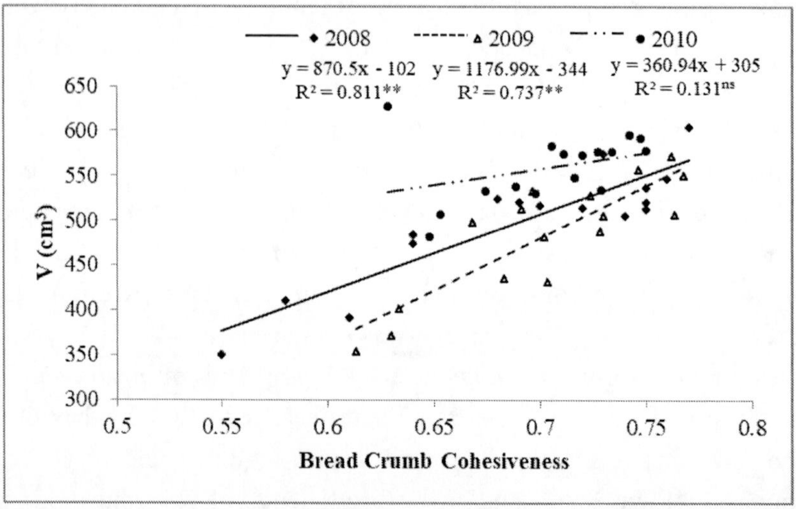

Figure 13. Linear regression relationship between bread volume and bread crumb cohesiveness in three different production seasons (* significant at 0.01 probability level; ** significant at 0.05 probability level; ns - not significant).

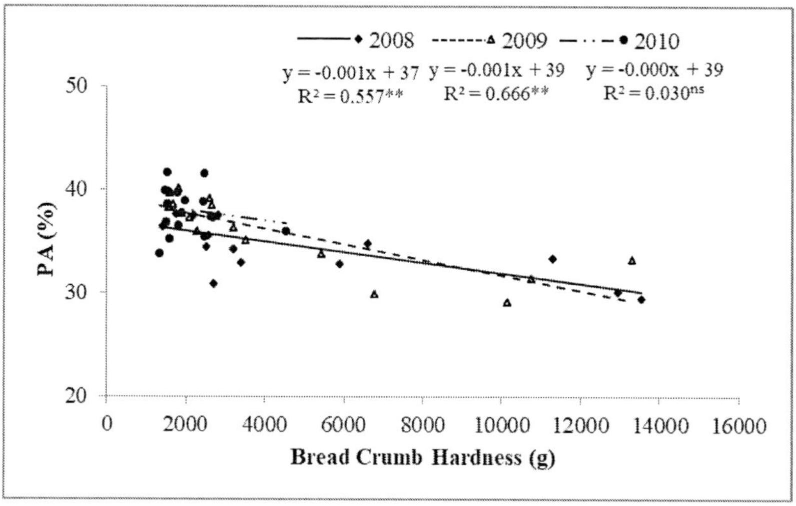

Figure 14. Linear regression relationship between of PA (%) and bread crumb hardness in three different production seasons (* significant at 0.01 probability level; ** significant at 0.05 probability level; ns - not significant).

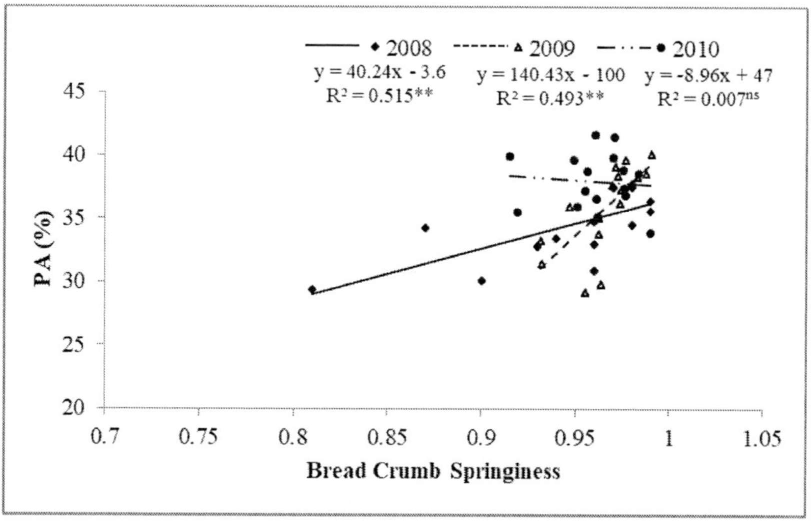

Figure 15. Linear regression relationship between of PA (%) and bread crumb springiness in three different production seasons (* significant at 0.01 probability level; ** significant at 0.05 probability level; ns - not significant).

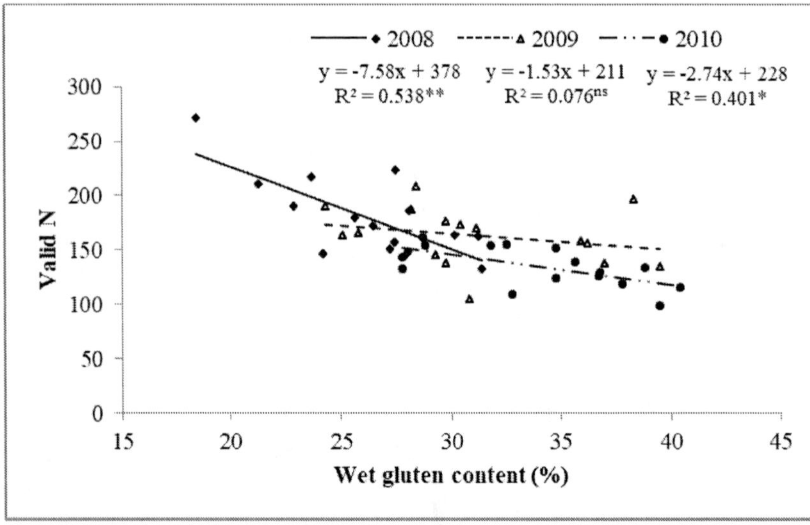

Figure 16. Linear regression relationship between of number of pores in bread crumb and wet gluten content in three different production seasons (* significant at 0.01 probability level; ** significant at 0.05 probability level; ns - not significant).

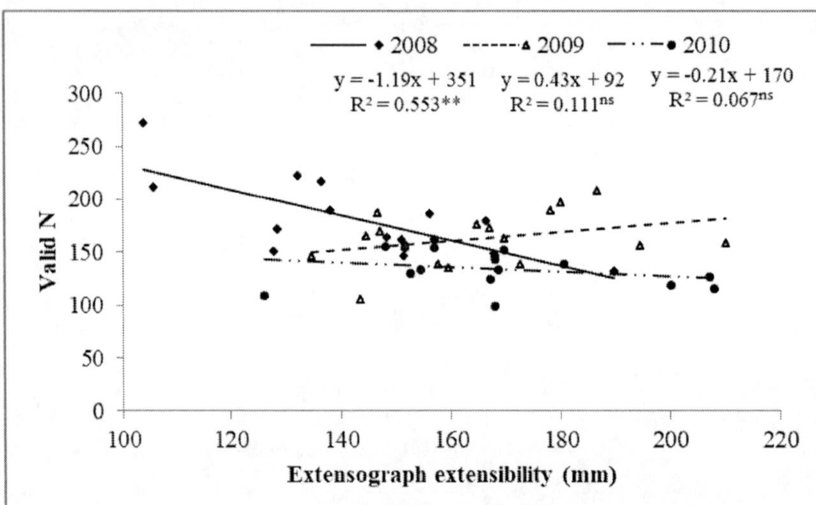

Figure 17. Linear regression relationship between of number of pores in bread crumb and bread extensograph extensibility in three different production season (* significant at 0.01 probability level; ** significant at 0.05 probability level).

Number of pores in bread crumb (Valid N) was negatively correlated with wet gluten content (Figure 16) in all three production seasons. Coefficients of determination of linear models in hot and wet seasons were statistically significant.

Extensograph extensibility showed negative correlation between Valid N in hot and wet seasons, whereas in the mixed season it was positive (Figure 17). Also, the linear regression model fit well only in the hot season due to coefficient of determination value.

Table 1 shows the statistically significant correlation coefficients ($P \leq 0.05$) for baking properties (bread volume, PA and Valid N) and most of important parameters of technological quality in the hot season.

Table 1. Correlation matrix (Pearson) of breadmaking quality parameters of wheat kernel in HOT season

Quality parameters	V (cm^3)	Valid N	PA (%)
FN (s)	ns	ns	ns
Starch damage	ns	ns	ns
L (mm)	**0.81**	**-0.70**	**0.69**
W (10^{-4}J)	**0.58**	ns	**0.53**
P/L	**-0.60**	**0.60**	ns
Energy (cm^2)	**0.56**	ns	**0.57**
Extensograph extensibility (mm)	**0.83**	**-0.74**	**0.73**
R/E	**-0.68**	**0.78**	ns
Wet gluten content (%)	**0.83**	**-0.73**	ns
Protein content (% d.m. basis)	**0.88**	**-0.77**	**0.51**
Hardness (g)	**-0.82**	**0.88**	**-0.75**
Springiness	**0.78**	**-0.78**	**0.72**
Cohesiveness	**0.90**	**-0.85**	**0.82**
Gumminess	**-0.80**	**0.87**	**-0.73**
Resilience	**0.75**	**-0.74**	**0.78**
V (cm^3)	**1**	**-0.85**	**0.73**
Valid N	ns	**1**	**-0.63**
PA (%)	ns	ns	**1**

Values in bold are different from 0 with a significance level alpha = 0.05, ns - not significant.

Significant coefficients of correlation were found among all of them except for a case FN and starch damage. As mentioned, various studies confirmed significant effect protein content wet, gluten content, extensograph and alveograph properties possess on bread volume (Bralard et al., 1991; Færgestad et al., 2000; Janssen et al., 1996; Příhoda, 1980; Ohm et al., 1998; Rubenthaler and Pomeranz, 1987; Tipples and Kilborn, 1974). Such influence of FN and starch damage on baking properties was expected. The weather conditions (amount of precipitation) that prevailed before and during the harvest did not allow preharvest sprouting and variation in FN and starch damage.

Table 2. Correlation matrix (Pearson) of breadmaking quality parameters of wheat kernel in the mixed season

Quality parameters	V (cm^3)	Valid N	PA (%)
FN (s)	ns	ns	ns
Starch damage	ns	ns	ns
L (mm)	**0.51**	ns	ns
W (10^{-4}J)	ns	ns	ns
P/L	ns	ns	ns
Energy (cm^2)	ns	ns	ns
Extensograph extensibility (mm)	ns	ns	ns
R/E	ns	ns	ns
Wet gluten content (%)	ns	ns	ns
Protein content (% d.m. basis)	ns	ns	ns
Hardness (g)	**-0.93**	**0.57**	**-0.82**
Springiness	**0.76**	**-0.63**	**0.70**
Cohesiveness	**0.86**	**-0.55**	**0.78**
Gumminess	**-0.94**	**0.59**	**-0.84**
Resilience	ns	ns	ns
V (cm^3)	**1**	**-0.52**	**0.82**
Valid N	ns	**1**	**-0.63**
PA (%)	ns	ns	**1**

Values in bold are different from 0 with a significance level alpha = 0.05, ns - not significant.

Table 3. Correlation matrix (Pearson) of breadmaking quality parameters of wheat kernel in the wet season

Quality parameters	V (cm^3)	Valid N	PA (%)
FN (s)	**-0.60**	ns	ns
Starch damage	ns	ns	ns
L (mm)	ns	ns	ns
W (10^{-4}J)	ns	ns	ns
P/L	ns	ns	ns
Energy (cm^2)	ns	ns	ns
Extensograph extensibility (mm)	ns	ns	ns
R/E	ns	ns	ns
Wet gluten content (%)	ns	**-0.63**	ns
Protein content (% d.m. basis)	ns	ns	ns
Hardness (g)	**-0.86**	ns	ns
Springiness	ns	ns	ns
Cohesiveness	ns	ns	ns
Gumminess	**-0.87**	ns	ns
Resilience	ns	ns	ns
V (cm^3)	**1**	ns	ns
Valid N	ns	**1**	ns
PA (%)	ns	ns	**1**

Values in bold are different from 0 with a significance level alpha = 0.05, ns - not significant.

Table 2 shows the statistically significant correlation coefficients ($P \leq 0.05$) for baking properties (bread volume, PA and Valid N) and most of important parameters of technological quality in the mixed season. Only parameters of TPA analysis, except for resilience, possessed significant coefficients of correlation with bread properties. Also, only L parameters were positively correlated with bread volume.

Table 3 shows the statistically significant correlation coefficients ($P \leq 0.05$) for relations among baking properties (bread volume, PA and Valid N) and most of important parameters of technological quality in hot season. Unexpectedly, in hot season only negative correlations exited between FN and bread volume; wet gluten content and valid N. The expected negative correlation was also found between hardness and bread volume and gumminess and bread volume.

Conclusion

The hot season (year with dry and hot weather from anthesis to full grain maturity) confirmed the well-known statistically significant correlation between protein content, wet gluten content, W, L, P/L (Alveograph properties), E, R/E (Exstensograph properties) and bread loaf volume. The same parameters were not statistically correlated with bread loaf volume in the mixed and wet seasons. Furthermore, out of all examined parameters of empirical rheology, only L showed positive correlation with bread volume in mixed season. Consequently, it could be concluded that the well-known relationship among bread volume and other breadmaking properties are not valid in the production seasons with increased precipitation from anthesis to full grain maturity (wet season).

Acknowledgments

This research was supported by the Ministry of Education, Science and Technological Development of the Republic of Serbia within the Project of technological development no. TR 31007 (2011–2018).

References

Addo, K., Coahran, D. R. and Pomeranz, Y. (1990). New Parameter Related to Loaf Volume Based on the First Derivative of the Alveograph Curve. *Cereal Chemistry,* 67: 64-69.

Aitken, T. R., and Geddes, W. F. (1934). The behaviour of strong flours of widely varying protein content when subjected to normal and severe baking procedures. *Cereal Chemistry,* 11: 487-504.

Aitken, T. R., and Geddes, W. F. (1939). The relation between protein content and strength of gluten–enriched flours. *Cereal Chemistry,* 16: 223-231.

Alcamo, J., Moreno, J. M., Nováky, B., Bindi, M., Corobov, R., Devoy, R. J. N., Giannakopoulos, C., Martin, E., Olesen, J. E. and Shvidenko, A. (2007) Europe. In: Parry, M. L., Canziani, O. F., Palutikof, J. P., van der Linden, P. J. and Hanson, C. E. (editors). *Climate Change 2007: Impacts, Adaptation and Vulnerability. Contribution of Working Group II to the Fourth Assessment Report of the Intergovernmental Panel on Climate Change.* Cambridge, UK: Cambridge University Press; 541-580.

Al-Saleh, A. and Brennan, C. S. (2012). Bread Wheat Quality: Some Physical, Chemical and Rheological Characteristics of Syrian and English Bread Wheat Samples. *Foods,* 1: 3-17.

Altenbach, S. B., DuPont, F. M., Kothari, K. M., Chan, R., Johnson, E. L., and Lieu, D. (2003). Temperature, water and fertilizer influence the timing of key events during grain development in a US spring wheat. *Journal of Cereal Science,* 37: 9-20.

American Association of Cereal Chemistry (2000). *Approved Methods of Analysis* (10th edn). *Methods* (44-15.02), (46-16.01) and (74–09). AACC, St. Paul, MN

Barak, S., Mudgil, D., and Khatkar, B. S. (2013). Relationship of gliadin and glutenin proteins with dough rheology, flour pasting and bread making performance of wheat varieties. *LWT – Food Science and Technology,* 51: 211-217.

Basso, B., Fiorentino, C., Cammarano, D., Cafiero, G., and Dardanelli, J. (2012). Analysis of rainfall distribution on spatial and temporal patterns of wheat yield in Mediterranean environment. *European Journal of Agronomy*, 41: 52–65.

Bocchiola, D., Nana, E. and Soncini, A. (2013). Impact of climate change scenarios on crop yield and water footprint of maize in the Po valley of Italy. *Agricultural Water Management,* 116: 50-61.

Borghi, B., Corbellini, M., Ciaffi, M., Lafiandra, D., De Stefanis, E., Sgrulleta, D., Boggini, G. and Di Fonzo, N. (1995). Effects of heat shock during grain filling on grain quality of bread and durum wheats. *Australian Journal of Agricultural Research,* 46: 1365-1380.

Borghi, B., Corbellini, M., Minoia, C., Palumbo, M., Di Fonzo, N., and Perenzin, M. (1997). Effects of Mediterranean climate on wheat bread–making quality. *European Journal of Agronomy,* 6: 145-154.

Branlard, G. Rousset, M., Loisel W. and Autran J. C. (1991). Comparison of 46 technological parameters used in breeding for bread wheat quality evaluation. *Journal of Genetics and Breeding*, 45: 263-280.

Carceller, J. L., and Aussenac, T. (2001). Size Characterisation of Glutenin Polymers by HPSEC–MALLS. *Journal of Cereal Science,* 33: 131-142.

Chopin, M. (1921). Relations entre les proprietes mecaniques des pates de farines et la panification [Relationship between the mechanical properties of pasta and breadmaking]. *Bulletin de la Societe d'Encouragement pour l'Industrie Nationale*, 133: 261.

Christensen, J. H. and Christensen O. B. (2007). A summary of PRUDENCE model projections of changes in European climate by the end of this century. *Climatic Change,* 81: 7-30.

Ciaffi, M., Tozzi, L., Borghi, B., Corbellini, M., and Lafiandra, D. (1996). Effect of Heat Shock during Grain Filling on the Gluten Protein Composition of Bread Wheat. *Journal of Cereal Science,* 24: 91-100.

Ćurić, D., Karlović, D., Tušak, D., Petrović B. and Đugum, J. (2001). Gluten as a Standard of Wheat Flour Quality. *Food Technology and Biotechnology,* 39: 353-361.

Czubaszek A., Subda H., Kowalska M., Korczak B., Żmijewski M. and Karolini-Skardzińska Z. (2001). Chemical and biochemical evaluation of chosen winter wheat variety flour. *Zywnosc: Nauka Technologia Jakosc*, 1: 77–84.

Czuchajowska Z. and Paszczynska B. (1996) Is Wet Gluten Good for Baking, *Cereal Chemistry,* 73: 483-489.

Daniel, C., and Triboi, E. (2000). Effects of Temperature and Nitrogen Nutrition on the Grain Composition of Winter Wheat: Effects on Gliadin Content and Composition. *Journal of Cereal Science,* 32: 45-56.

Debaeke, P., Aussenac, T., Fabre, J. L., Hilaire, A., Pujol, B., and Thuries, L. (1996). Grain nitrogen content of winter bread wheat (*Triticum aestivum L.*) as related to crop management and to the previous crop. *European Journal of Agronomy,* 5: 273-286.

Dobraszczyk, B. J. (1997). Development of a new dough inflation system to evaluate doughs. *Cereal Foods World,* 42: 516-519.

Dobraszczyk, B. J., & Roberts, C. A. (1994). Strain hardening and dough gas cell–wall failure in biaxial extension. *Journal of Cereal Science,* 20: 265-274.

Dobraszczyk, B. J., and Morgenstern, M. P. (2003). Rheology and the breadmaking process. *Journal of Cereal Science,* 38: 229-245.

Don, C., Lookhart, G., Naeem, H., MacRitchie, F., and Hamer, R. J. (2005). Heat stress and genotype affect the glutenin particles of the glutenin macropolymer–gel fraction. *Journal of Cereal Science,* 42: 69-80.

Dowell, F. E., Maghirang, E. B., Pierce, R. O., Lookhart, G. L., Bean, S. R., Xie, F., Caley, M. S., Wilson, J. D., Seabourn, B. W., Ram, M. S., Park, S. H. and Chung, O. K. (2008). Relationship of bread quality to kernel, flour, and dough properties. *Cereal Chemistry,* 85: 82-91.

Dumur, J., Jahier, J., Dardevet, M., Chiron, H., Tanguy, A. M., and Branlard, G. (2010). Effects of the replacement of Glu–A1 by Glu–D1 locus on agronomic performance and bread–making quality of the hexaploid wheat cv. Courtot. *Journal of Cereal Science,* 51: 175-181.

DuPont, F. M., Altenbach, S., Chan, R., Cronin, K., and Lieu, D. Interactions between fertilizer, temperature and drought in determining flour composition and quality for bread wheat (2000). In: Shewry, P. R. and Tatham A. S. editors. *Wheat gluten.* Cambridge, UK: Gluten Royal Society of Chemistry, 488-491.

Færgestad E. M., Molteberg E. L. and Magnus M. (2000). Interrelationships of protein composition, protein level, baking process and the characteristics of hearth bread and pan bread. *Journal of Cereal Science,* 31: 309-320.

Garrido–Lestache, E., López–Bellido, R. J., and López–Bellido, L. (2004). Effect of N rate, timing and splitting and N type on bread–making quality in hard red spring wheat under rainfed Mediterranean conditions. *Field Crops Research,* 85: 213-236.

Gibson, L. R., and Paulsen, G. M. (1999). Yield components of wheat grown under high temperature stress during reproductive growth. *Crop Science,* 39: 1841-1846.

Gil, D. H., Bonfil, D. J., and Svoray, T. (2011). Multi scale analysis of the factors influencing wheat quality as determined by Gluten Index. *Field Crops Research,* 123: 1-9.

Giorgi, F., Bi, X. and Pal, J. (2004). Mean, interannual and trends in a regional climate change experiment over Europe, II: climate change scenarios (2071–2100). *Climate Dynamics.* 23: 839-858.

Gooding, M. J., Ellis, R. H., Shewry, P. R., and Schofield, J. D. (2003). Effects of Restricted Water Availability and Increased Temperature on the Grain Filling, Drying and Quality of Winter Wheat. *Journal of Cereal Science,* 37: 295-309.

He, W. T., Yang, J. Y., Drury, C. F., Smith, W. N., Grant, B. B., Qian, B., Zhou, W. and Hoogenboom G. (2018). Estimating the impacts of climate change on crop yields and N_2O emissions for conventional and no-tillage in Southwestern Ontario, Canada. *Agricultural Systems,* 159: 187-198.

Högy, P., and Fangmeier, A. (2008). Effects of elevated atmospheric CO_2 on grain quality of wheat. *Journal of Cereal Science,* 48: 580-591.

Hungarian Standard (1988). Lisztvizsgálati módszerek. Vízfelvevő képesség és sütőipari érték [Flour testing methods. Determination of water absorption capacity and baking quality]. *Method* MSZ 6369/6, MSZT, Budapest.

Indrani, D., Swetha, P., Soumya, C., Rajiv, J., and Venkateswara Rao, G. (2011). Effect of multigrains on rheological, microstructural and quality characteristics of north Indian parotta – An Indian flat bread. *LWT – Food Science and Technology,* 44: 719-724.

Indrani, D., Swetha, P., Soumya, C., Rajiv, J., and Venkateswara Rao, G. (2011). Effect of multigrains on rheological, microstructural and quality characteristics of north Indian parotta – An Indian flat bread. *LWT – Food Science and Technology,* 44: 719-724.

International Association for Cereal Science and Technology (2006). Standard Methods of the International Association for Cereal Science and Technology (6th edn). *Methods* (114/1), (121), (106/2) and (107/1), ICC, Vienna

Janssen A. M., Van Vliet T. and Vereiken J. M. (1996). Fundamental and empirical rheological behaviour of wheat flour doughs and comparison with bread making performance. *Journal of Cereal Science,* 23: 43-54.

Johnson, V. A., and Mattern, P. J. (1987). Wheat, rye and triticale. In: Olson R. A. and Frey K. J. editors. *Nutritional Quality of Cereal Grains: Genetic and Agronomy Improvements.* Madison, WI, USA: American Society of Agronomy, Inc; 133-182.

Jones, P. D. and Moberg, A. (2003). Hemispheric and large scale surface air temperature variations: an extensive revision and an update to 2001. *Journal of Climate,* 16: 206-223.

Kaluđerski, G. and Filipović, N. (1998). Laboratorijsko mlevenje pšenice [Wheat milling in Laboratories]. In: Žeželj, M. and Vukobratović, R. editors. *Metode ispitivanja kvaliteta žita, brašna i gotovih proizvoda* [*Methods of testing the quality of wheat, flour and end products*]. Novi Sad, SRJ Jugoslavija: Tehnološki fakultet u Novom Sadu, Zavod za tehnologiju žita i brašna; 62-70.

Konopka, I., Fornal, Ł., Abramczyk, D. Rothkaehl, J. and Rotkiewicz, D. (2004). Statistical evaluation of different technological and rheological tests of Polish wheat varieties for bread volume prediction. *International Journal of Food Science and Technology,* 39: 11-20.

Ktenioudaki, A., Butler, F., and Gallagher, E. (2011). Dough characteristics of Irish wheat varieties II. Aeration profile and baking quality. *LWT – Food Science and Technology,* 44: 602-610.

Lansigan, F. P., de los Santos, W. L. and Coladilla, J. O. (2000). Agronomic impacts of climate variability on rice production in the Philippines. Agriculture. *Ecosystems and Environment,* 82: 129-137.

Launay, B., Bure, J., and Praden, J. (1977). Use of the Chopin Alveograph as a rheological tool. I. Dough deformation measurements. *Cereal Chemistry,* 54: 1042-1048.

Liu, D. L., Zeleke, K. T., Wang, B., Macadam, I., Scott, F. and Martin, R. J. (2017). Crop residue incorporation can mitigate negative climate change impacts on crop yield and improve water use efficiency in a semiarid environment. *European Journal of Agronomy,* 85: 51-68.

Lobell, D. B. and Gourdji, S. M. (2012). The influence of climate change on global crop productivity. *Plant Physiology,* 160: 1686-1697.

López–Bellido, L., López–Bellido, R. J., Castillo, J. E., and López–Bellido, F. J. (2001). Effects of long–term tillage, crop rotation and nitrogen fertilization on bread–making quality of hard red spring wheat. *Field Crops Research,* 72, 197-210.

Mastilović, J., Živančev, D., Lončar, E., Malbaša, R., Hristov, N. and Kevrešan, Ž. (2018). Effects of high temperatures and drought during anthesis and grain filling period on wheat processing quality and underlying gluten structural changes, *Journal of the Science of Food and Agriculture,* 98: 2898-2907:

Miś, A., and Dziki, D. (2013). Extensograph curve profile model used for characterising the impact of dietary fibre on wheat dough. *Journal of Cereal Science,* 57: 471-479.

Ohm J. B., Chung O. K. ans Deoye C. W. (1998). SKCS of HRW wheat in relation to milling and baking quality. *Cereal Chemistry,* 75: 151-161.

Pandey, R. (2017). Farmers' perception on agro-ecological implications of climate change in the Middle-Mountains of Nepal: a case of Lumle Village, Kaski. *Environment, Development and Sustainability*, 1-27 https://doi.org/10.1007/s10668-017-0031-9.

Panozzo, J. F., and Eagles, H. E. (2000). Cultivar and envirnomental effects on quality characters in wheat. II. Protein. *Australian Journal of Agricultural Research,* 51: 629-636.

Patil, S. D., Preethi, B., Bansod, S. D., Singh, H. N., Revadekar, J. V. and Munot A. A. (2013). On the association between pre-monsoon aerosol and all-Indian summer monsoon rainfall. *Journal of Atmospheric and Solar-Terrestrial Physics,* 102: 1-7.

Pena, R. J., Amaya, A., Rajaram, S., and Mujeeb–Kazi, A. (1990). Variation in quality characteristics associated with some spring 1B/1R translocation wheats. *Journal of Cereal Science,* 12: 105-112.

Peterson, C. J., Graybosch, R. A., Shelton, D. R., and Baenziger, P. S. (1998). Baking quality of hard winter wheat: Response of cultivars to environment in the Great Plains. *Euphytica,* 100: 157-162.

Picchi, V., Iriti, M., Quaroni, S., Saracchi, M., Viola, P. and Faoro, F. (2010). Climate Variations and Phenological Stages Modulate Ozone Damages in Field-Grown Wheat. A Three-Year Study with Eight Modern Cultivars in Po Valley (Northern Italy). *Agriculture Ecosystems and Environment,* 135: 310-317.

Pomeranz, Y., Peterson, C. J., and Mattern, P. J. (1985). Hardness of winter wheats grown under widely different climatic conditions. *Cereal Chemistry,* 62: 463-467.

Pomeranz, Y., Shogren, M. D., Finney, K. F., and Bechtel, D. B. (1977). Fiber in breadmaking—effects on functional properties. *Cereal Chemistry,* 54: 25-41.

Preston, K. R., Kilborn, R. H., and Dexter, J. E. (1987). Effects of starch damage and water absorption on the alveograph properties of Canadian hard red spring wheats. *Canadian Institute of Food Science and Technology Journal,* 20: 75-80.

Příhoda J. (1980). Víceleté sledování korelací mezi vlastnomosti zrna, mouky a pečených výrobků [Multi-year tracking correlations between grain, flour, and baking properties]. *Research report, VÚMPP*, Praha: 26-85.

Rakszegi, M., Pastori, G., Jones, H. D., Békés, F., Butow, B., Láng, L., Bedő, Z. and Shewry, P. R. (2008). Technological quality of field grown transgenic lines of commercial wheat cultivars expressing the 1Ax1 HMW glutenin subunit gene. *Journal of Cereal Science,* 47: 310-321.

Różyło R. and Laskowski J. (2011). Predicting Bread Quality (Bread Loaf Volume and Crumb Texture). *Polish Journal of Food and Nutrition Sciences,* 61: 61-67.

Rubenthaler G. L. and Pomeranz Y. (1987). NIR spectra of HRW wheat varyng widely in protein content and bread potencial. *Cereal Chemistry,* 64: 407-411.

Ruml, M., Vuković, A., Vujadinović, M., Djurdjević, V., Ranković–Vasić, Z., Atanacković, Z., Sivčev, B., Marković, N., Matijašević S. and Petrović, N. (2012). On the use of regional climate models: Implications of climate change for viticulture in Serbia. *Agricultural and Forest Meteorology,* 158-159: 53-62.

Schuler, S. F., Bacon, R. K., Finney, P. L., and Gbur, E. E. (1995). Relationship of test weight and kernel properties to milling and baking quality in soft red winter wheat. *Crop Science,* 35: 949-963.

Shrestha, S., Chapagain, R. and Babel, M. S. (2017). Quantifying the impact of climate change on crop yield and water footprint of rice in the Nam Oon Irrigation Project. *Thailand, Science of the Total Environment,* 599-600: 689-699.

Skendi, A., Biliaderis, C. G., Papageorgiou, M., and Izydorczyk, M. S. (2010). Effects of two barley β–glucan isolates on wheat flour dough and bread properties. *Food Chemistry,* 119: 1159-1167.

Spiertz, J. H. J., Hamer, R. J., Xu, H., Primo–Martin, C., Don, C., and van der Putten, P. E. L. (2006). Heat stress in wheat (*Triticum aestivum L.*): Effects on grain growth and quality traits. *European Journal of Agronomy,* 25: 89-95.

Stone, P. and Savin, R. (1999). Grain quality and its physiological determinants. In: Satorre E. H. and Slafer G. A. editors. *Wheat Ecology and Physiology of Yield Determination.* New York; USA: Food Products Press/The Haworth Press; 85-120.

Tashiro, T., and Wardlaw, I. F. (1990). The response to high temperature shock and humidity changes prior to and during the early stages of grain development in wheat. *Australian Journal of Plant Physiology,* 17: 551-561.

Tipples, K. H. and Kilborn R. H. (1974). Baking strength index and relation of protein conetnt to bread volume. *Canadian Journal of Plant Science,* 54: 231-234.

Webber, H. Zhao, G., Wolf, J., Britz, W., de Vries, W., Gaiser, T., Hoffman, H. and Ewert, F. (2015). Climate change impacts on European crop yields: Do we need to consider nitrogen limitation? *European Journal of Agronomy,* 71: 123-134.

Wei, T., Cherry, T. L., Glomrød, S. and Zhang, T. (2014). Climate change impacts on crop yield: Evidence from China. *Science of the Total Environment,* 499: 133-140.

Zhang, P., He, Z., Chen, D., Zhang, Y., Larroque, O. R., and Xia, X. (2007). Contribution of common wheat protein fractions to dough properties and quality of northern–style Chinese steamed bread. *Journal of Cereal Science,* 46: 1-10.

Zhao, C. X., He, M. R., Wang, Z. L., Wang, Y. F., and Lin, Q. (2009). Effects of different water availability at post–anthesis stage on grain nutrition and quality in strong–gluten winter wheat. *Comptes Rendus Biologies,* 332: 759-764.

Zimmermann, A., Webber, H., Zhao, G., Ewert, F., Kros, J., Wolf, J., Britz, W. and de Vries, W. (2017). Climate change impacts on crop yields, land use and environment in response to crop sowing dates and thermal time requirements. *Agricultural Systems,* 157: 81-92.

In: Agricultural Research Updates. Volume 27 ISBN: 978-1-53615-916-5
Editors: Prathamesh Gorawala et al. © 2019 Nova Science Publishers, Inc.

Chapter 7

ORGANIC VS. CONVENTIONAL FARMING OF RASPBERRIES: CHEMICAL AND SENSORY CHARACTERISTICS – A REVIEW

Alice Vilela[1], Teresa Pinto[2], Rosário Anjos[2] and Fernanda Cosme[1,*]

[1]CQ-VR, Chemistry Research Centre,
University of Trás-os-Montes and Alto Douro, ECVA-DeBA,
Vila Real, Portugal
[2]CITAB, Centre for the Research and Technology
of Agro-Environmental and Biological Sciences,
University of Trás-os-Montes e Alto Douro, ECVA-DeBA,
Vila Real, Portugal

ABSTRACT

Organic farming is distinguished from other production systems because it excludes almost all synthetic chemicals, uses cultural rotations,

* Corresponding Author's E-mail: fcosme@utad.pt.

crop residues, animal manure, and all organic waste from the farm. Several studies have demonstrated that fruits and vegetables from organic farming are generally endowed with enhanced physicochemical and sensory properties. Red fruits should be part of a balanced diet due to their characteristic and specific content of phenolic compounds, consisting essentially of anthocyanins, flavonols, proanthocyanidins and phenolic acids, known for their high antioxidant capacity. However, the studies described by different authors are inconsistent when indicating that organic and conventional farming of fruits and vegetables differs concerning physicochemical and sensory characteristics. So, the aim of this chapter is to summarize recent knowledge in order to contribute to a better understanding of how farming practices affect berry composition and consequently, the sensory characteristics of raspberries.

Keywords: organic *vs.* conventional farming, raspberry, chemical composition, sensory profile, consumer's choice

INTRODUCTION

Consumers' eating habits are changing as a result of increased awareness regarding their health. It is widely believed that organic farming products are more abundant in compounds necessary to maintain adequate health. An increasing number of studies are being conducted comparing the chemical composition of organic *vs.* conventional farming products and their impact on human health and environment [1]. Even so, there is a great lack of knowledge about the composition differences of organic fruits such as raspberry.

This reality has altered the interest of both private and governmental entities in investing in agriculture, not in intensive agriculture practices without respect for the environment, but in agriculture practices that are safer for the environment and for the health of people and animals [2]. Agricultural practices have the potential to improve the increment of most nutrients. This effect depends on the levels and forms of nitrogen applied to the crops, the balance of macro- and micronutrients in the soil and on the soil quality [3]. Thus, over the last few years a significant increase in the investigation and discussion of the duality of organic *vs.* conventional farming can be perceived.

PRACTICAL ASPECTS OF ORGANIC VS. CONVENTIONAL FARMING

The organic *vs.* conventional farming dichotomy has never been as debated as it is currently because consumers have never been concerned about the consumption of healthy but safe food for the environment as they are now. Indeed, there is a worldwide growing demand for healthy and sustainable food. Furthermore, the development of sustainable food systems is a growing concern among a number of intergovernmental organizations [4, 5].

The definition of food quality is constantly changing. Figure 1 summarizes ecological criteria of food quality according to Vogtmann [6]. The definition of quality products was improved in 2005 by Luning et al. [7], where physical, chemical and external characteristics of the product appear connected with production characteristics, environmental aspects, and marketing. Consequently, ecological and holistic criteria should be used when considering the food production chain and food quality.

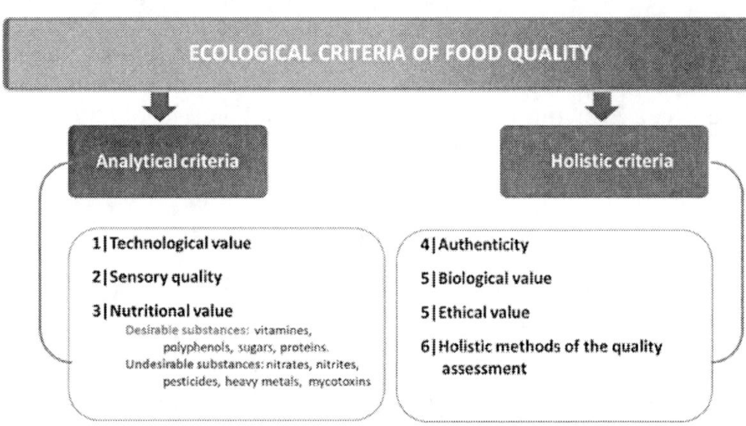

Figure 1. Ecological criteria of food quality. Adapted from Vogtmann [6].

According to FAO [8], about 40% of the land is dedicated to agriculture with a huge environmental impact. Conventional farming practice in an intensive way may have a negative impact on the environment, essentially due to the high inputs of synthetic pesticides and mineral fertilizers, chemical protection treatments and highly water dependent [9]. The maximum production with the minimum costs is the principle of conventional agriculture. On the other hand, according to European Union (EU) Regulation EC Nº 834/2007, organic farming practice is defined as

> 'an overall system of farm management and food production that combines best environmental practices, a high level of biodiversity, the preservation of natural resources, the application of high animal welfare standards and a production method in line with the preference of certain consumers for products produced using natural substances and processes.'

The concept of organic farming is based on a number of different principles and practices, aiming to reduce the human impact on the environment and to ensure that the agricultural system functions as naturally as possible, with a focus on health and well-being. In this way, the organic agriculture requires the use of organic fertilizers, non-chemical measures for pest control and prevention and livestock access to the outdoors and roughage feed [2]. It is proving to be a promising type of production to meet the challenges of modern agriculture. Throughout the world, there is a growing of agricultural area devoted to organic farming, either by the installation of new agricultural parcels entirely dedicated to organic production or by converting existing areas, being over 50.9 million hectares recorded in 2015 [10]. Taking the European Union as an example, its evolution over the last thirty years is 0.1%, 0.6%, 3.6%, and 6.2% of agricultural area for organic farming in 1985, 1995, 2005, and 2015, respectively, representing 11.2 million hectares in 2015 [8, 10, 11].

The area for the temperate fruit production has followed this growth. According to Willer et al. [10] it was of about 12.5 million hectares in 2013 worldwide. Hence, the producers seek to respond to the demand of consumers who are increasingly seeking fruit free from biotic and xenobiotic contaminants but with high nutritional, sensory and nutraceutical characteristics.

Cultivation method, genotype and environment interactions are the most import factors that affect the sensory properties and phytochemical composition of plants and fruits [12]. In organic farming, green manure, animal fertilizers, a varied crop rotation [13], agrobiodiversity, cover crops, mulching, incorporation of nitrogen fixation fertilizers from plants and other biodegradable inputs are favored [14]. Herbicides and pesticides are not allowed [15], so natural plant defense mechanisms such as secondary metabolites, increasing crop genetic diversity, enhance pollination services, against pests and diseases are activated [16]. Weed control is based on the selection of varieties and mechanical methods [13].

Effect of Organic *vs.* Conventional Farming on Raspberry Chemical Composition

Organic farming systems are increasing in terms of production area as well as in the number of operators. The main goal includes a rational use of natural properties with high quality and shelf life performance. More in-depth studies may relate the organic *vs* conventional farming comparison to the more general issue of pre-harvest effects on post-harvest crops performance [17]. The balance between primary and secondary metabolic pathways appears to be an essential aspect resulting from the complex interaction of genotype, environment, and agricultural practices leading to modifications in quality and post-harvest performance of fresh fruits and vegetables. The necessity to increase food quality, available in the world in a defensible way that should orient research on agricultural practices in order to raise the nutritional composition of fresh fruits and vegetables and to enhance shelf life [17].

Organic farming enhances soil fertility by the addition of nutrient adjustments excluding synthetic materials such as mineral inorganic fertilizers, leading health benefits [18]. The consumers of organic farming products assume that their nutritional properties are a major benefit as well as the effects on the environment [19]. It was found that the biodiversity of plants, invertebrates, and birds were higher on organic farms compared to conventional farming practices [20]. So, the most important of organic farming is the removal of the use of synthetically based products which might harm soil productivity, the ecosystem, and the groundwater [21].

Red raspberry (*Rubus ideaus* L.) is an important fruit with potential health benefits that growths in cold and temperate regions of the world [22]. According to several studies [23, 24, 25], the interest in raspberry production in open, high tunnel and greenhouse conditions has increased. However, the changing climatic conditions are becoming a limiting factor in the production of berry fruit from existing commercial cultivars in the North Sea Region [26].

While raspberries are conventionally produced in the field, new protected production systems are currently established. These new production systems need to protect productivity during climate changes problems to increase productivity. These problems need to be addressed to sustain the raspberry industry in the future [27].

Berry fruits in general and particularly raspberry have been collected and consumed worldwide for a long time, since they were recognised for their possible health benefits and due to their natural phytochemicals as well as their taste. They are popular for its flavor, nutritious and attractive red color as a source of many bioactive substances, containing high levels of phenolic compounds (anthocyanins), phenolic acids, ellagitannins, dietary fibre [28, 29], α-tocopherol [30], tocotrienol, calcium, potassium, magnesium, carotenoids, linoleic acid and vitamin C [30, 31, 32, 33, 34] being considered a good source of natural antioxidants, which could be considered a good source to provide protection against various human diseases caused by oxidative stress [35, 36].

Raspberry is a good example that fruits from organic farming contain a higher level of some phytochemicals and antioxidant capacities than those produced by conventional farming [37]. These compounds such as flavonoids are natural antioxidants that have an importance in the prevention of cardiovascular diseases and cancer, as well as contributing together with vitamin C to the strengthening of blood vessels [38]. In addition, anthocyanins showed high influence in lowering mortality caused by heart disease. Anthocyanins are characterized by anti-inflammatory, antibacterial activity, and also beneficial effect on the eyeball improving the ability to see, as well as increase the elasticity of the blood vessels [39]. The high concentration of phenolics acids is essential to health due to their anti-atherosclerotic, antitumor and antimicrobial features. They also have properties to remove reactive oxygen species (ROS), scavenge free radicals, causing the chelation of metal ions and the inhibition of enzymes from the group of oxidases [36].

The large scale of agricultural chemicals used in conventional farming food production makes consumers search for wholesome, healthy and free from contamination products on the market. This increases the choice for food from organic farming considered to have better quality and being healthier [36].

Conventional farming represents a more sustainable strategy than organic farming at a global level in order to meet the needs of continuous population expansion. However, there is a valid concern about the negative impacts of conventional farming on the environment and health that need to be discussed and many harmful practices should be changed. Globally, the implications for food security must be assessed in the choice of conventional *vs.* organic farming [40].

Vitamins and Minerals

In raspberries vitamin C ranged from 15.6 to 54.2 mg/100g fresh weight [41, 42, 43, 44, 45, 46]. Therefore, raspberry is also a good source of vitamin C, an antioxidant with anticancer and immunomodulatory properties [43, 47, 48, 49].

Raspberreies contain mineral such as calcium [29, 50], potassium [29, 50] and magnesium [50] (Table 1).

Table 1. Metal contents in raspberry fruits [50]

	mg/100 g dried weight
Phosphate	741.97 ± 86.56
Iron	9.06 ± 2.60
Manganese	1.74 ± 0.77
Zinc	2.97 ± 0.17
Calcium	200.14 ± 40.66
Potassium	1123.70 ± 216.29
Magnesium	159.28 ± 17.73
Sodium	4.70 ± 1.63

Sugars and Organic Acids

The main sugars presented in raspberries from different regions or cultivated under conventional as well as organic farming were fructose 35-45%, glucose 30-35% and, sucrose 30-35% [42, 51]. The total sugars ranged from 2.07 to 7.70 g/100g fresh weight [42, 51]. Total acidity in raspberries ranged from 1.5-3.7 (%) [44, 45, 52]. Raspberries pH ranged from 2.60 to 3.50 [45, 52].

Phenolic Compounds

Berry fruits comprise a variety of phenolic compounds such as phenolic acids, flavonoids, lignans and polymeric tannins placed in the plant tissues, frequently in the surface layer of the plant or berry fruit [34]. The total phenolic compounds in raspberries range from 186.9 to 714.7 mg/100 g of fresh weight [45, 51]. Anthocyanins constitute the main group of phenolic compounds in berries. Normally, the concentration of flavonols is higher than that of flavan-3-ols in berries. Nevertheless, in raspberry, the concentration of flavan-3-ols has been described to be higher than that of flavonols by several researchers [53, 54]. Therefore, the main phenolic compounds presented in raspberries are flavonoids such as anthocyanins (colored substances), ellagic acid and ellagitannins and phenolic polymer

[34]. There are few studies that study the concentration of phenolic acids in berries, although it is higher than that of flavonols or flavan-3-ols in raspberry (Table 2). In raspberry, the concentration of ellagic acid is described to be high [55], but not higher than that of anthocyanins (Table 2).

Raspberries contain sanguiin H-6 as the main ellagitannin [45, 64, 65] and the tetramer lambertianin C [45, 65, 66].

Ellagic acid is the predominant phenolic compound in raspberries present in free form or esterified to glucose [45, 65, 67, 68] in a concentration ranging from 79.0 to 246.3 mg/100 g of fresh weight [69] and it is released after acid hydrolysis of ellagitannins that is present in an average concentration of 30 mg/100 g of fresh weight, and as referred previously mainly in the form of sanguiin H-6 and lambertianin [49, 67]. Ellagic acid has acknowledged more consideration for its nutritional and pharmacological potential as it has been described to have antioxidant and antiviral activity [70]. Furthermore, it has also been established that ellagic acid may act as a chemopreventive agent against several types of cancer [71]. The main phenolic acids are hydroxycinnamic or hydroxybenzoic acid derivatives [34]. The hydroxybenzoic acids are presented in both free and esterified form and consist of *p*-hydroxybenzoic acid, salicylic acid, gallic acid, and ellagic acid. Also, hydroxybenzoic acids are constituents of complex structures such as hydrolyzable ellagitannins [72, 73].

Table 2. Flavonoid and phenolic acid concentration in raspberries

Flavonoids	mg/kg fresh weight	Reference
Kaempferol	284–1240	[55, 56]
Quercetin	8–29	[57, 58]
Total anthocyanins	230–9950	[54, 59, 60, 48]
Total flavan-3-ols	32–480	[54, 61]
Phenolic acids		
p-Coumaric acid	6–25	[62, 63]
Caffeic acid	4–10	[62, 63]
Ferulic acid	3–17	[62, 63]
Gallic acid	15–59	[58, 63]
Ellagic acid	19–244	[46, 63]

Table 3. Concentration of proanthocyanidins in raspberries [78]

	mg/100 g (fresh weight)
Monomers	4.4 ± 3.4
Dimers	11.5 ± 9.8
Trimers	5.7 ± 5.5
4–6 degree of polymerization	7.7 ± 5.2
7–10 degree of polymerization	0.9 ± 2.2
Total proanthocyanidins	30.2 ± 23.4

Raspberries contained 7.7 µg/g of total flavonols with 86% corresponding to quercetin, kaempferol has also been detected in raspberries [67, 74].

Cyanidin is the most common anthocyanidin in raspberries, being the three main anthocyanins cyanidin-3-sophoroside, cyanidin-3-glucosyrutinoside and cyanidin-3-glucoside [49, 75, 76, 77]. Total anthocyanins raged from 2.1 to 400 mg/100 g fresh weight [43].

Raspberries contain 0.3 mg/g polymeric procyanidins according to Gu et al. [78] as shown in Table 3.

The differences in phenolic compounds composition observed could be due to the growing conditions, such as soil, geographical and environmental conditions during the raspberry fruit development, degree of maturity at harvest and also related to genetic differences [79]. For example caffeic acid, *p*-coumaric acid, and ferulic acid content are normally higher in young raspberries fruits decreasing first fast and then more slowly during fruit maturation and postharvest storage [63]. The anthocyanins and total phenolics concentration increase with raspberry fruit maturity [60]. Furthermore, the concentration of flavan-3-ols declines throughout raspberries fruit growth and maturation [63].

Aroma and Flavor Compounds

"Aroma" denotes a compound that is volatile and responsible for an odor, which can have pleasant or unpleasant connotations. It is the sensation perceived by the nose when sniffing, when the air goes to the ortho-nasal passage. The word "flavor" mentions the awareness resulting from stimulating, at the same time as the taste buds (in the tongue) and the olfactory receptors (in the nose). The air passes through the retro-nasal route.

As to other fruits, the volatile composition of raspberries is influenced by numerous factors, from cultivar variation to soil composition, climate and fruit ripeness [80]. Some raspberry volatiles are significant not only for the perception of fruit sensory quality but, also, for the resistance to molds [81], such as the case of monoterpenes [82].

Raspberry present a huge amount of volatile compounds, a summary of several of them, referred to in the literature since 1962 [83], is presented in Table 4.

Acids appear cited and quantified since 1976 [84], and possess a very unpleasant smell. The majority of the descriptors used for acids characterization is "stinky cheese," "smelly," "sour milk," "rancid oil," among others.

Alcohols are present on a wide range of concentrations (from 0.0007 to 3.5 mg/kg), being mentioned in 1962, in the publication of Winter and Sundt [83]. The alcohols possess pleasant odor descriptors, from fruits to flowers: fruity, melon, violet, herbal, peony, and rose, orange, citrus, orange and lavender; and very unpleasant ones: fuel, phenolic, cheesy, musty, and waxy.

Aldehydes are also present in a huge range of concentrations, being acetaldehyde the one that presented the highest concentration (26 mg/kg) in the pioneering work of Winter and Sundt [83] and possessing a pungent, ethereal, aldehydic and fruity aroma.

In the ketones family, Honkanen and co-workers, in 1980 [85], identified and quantified the 6-methyl-5-hepten-2-one, which presents a green, citrus, musty, lemongrass or apple aroma. This compound was detected at values higher than 3.0 mg/kg. Other ketones, occurring in vestigial amounts, have also been identified: 2,3-butanedione (diacetyl), 2-heptanone, 2-nonanone, 3-hydroxybutanone (acetoin), propan-2-one (acetone) and acetophenone [69, 83, 84, 86, 87, 88].

Table 4. Volatile compounds in raspberry fruit (*Rubus idaeus* L.). Reported minimum and maximum values (mg/kg) and associated odor description

Compounds		Quantitative data (minimum and maximum value mg/kg)	Olfactory descriptor TGSC Information System [93]	References
Acids				
1	Isopentanoic acid (3-methylbutanoic acid)	0.05	Rancid, vomit, sweaty feet, wet goat or stinky cheese.	[84]
2	Acetic acid	0.0205–0.275; 5.4–135	Vinegar and acid taste.	[69, 84, 90, 91]
3	Butanoic acid	0.25; 0.6	Sick, smelly feet, sweat, gone-off curry, sour milk.	[84, 85]
4	Heptanoic acid	0.15	Rancid, sour, cheesy, sweat.	[84]
5	Hexanoic acid	0.0286–0.1586; 6.7	Sour, fatty, sweat, cheese.	[69, 84, 85, 91]
6	Nonanoic acid	0.05	Waxy, dirty, cheese cultured, dairy.	[84]
7	Octanoic acid	0.6	Fatty, waxy, rancid oily, vegetable, cheesy.	[84, 85]
Alcohols				
8	(Z)-3-Hexen-1-ol octanoic acid	0.0091–0.01500; 0.06–0.47	Green like grass, fruity, cheesy, musty, waxy, ripe, heavy, and slightly buttery.	[69, 84, 85, 86, 87, 88, 91, 94]
9	(Z)-Octen-2-ol	0.005–0.05	Creamy, mushroom, melon, rind, waxy	[84]
10	1-Butanol	0.05; 0.01–0.1	Roasted wine, onion, fruity, fuel, alcoholic, whiskey.	[84, 93]
11	1-Heptanol	0.005–0.05	Musty, leafy, violet, herbal, green, sweet, wood, peony.	[86]
12	1-Hexanol	0.0008–0.0029; 1.4	Ethereal, fusel oil, fruity, alcoholic, sweet, green.	[69, 84, 85, 86, 93]
13	1-Nonanol	0.005–0.05	Fresh clean, fatty, floral, rose, orange, dusty, wet, oily.	[86]
14	1-Octanol	0.005–0.05	Waxy, green, orange, aldehydic, rose, mushroom.	[86]
15	1-Pentanol	<0.01; 0.005–0.05	Fusel, cognac, wine, cocoa, green, fruity.	[85, 86, 93]
16	2-Heptanol	0.00073–0.00664	Fresh, lemon, grass, herbal, sweet, floral, fruity, green.	[69]
17	2-Nonanol	0.002–0.012; 0.003–0.007	Waxy, green, creamy, citrus, orange, cheese, fruity.	[87, 88]

Compounds		Quantitative data (minimum and maximum value mg/kg)	Olfactory descriptor TGSC Information System [93]	References
18	2-Phenylethanol	0.5	Floral, rose, dried rose, flower, rose water.	[85]
19	3-Methyl-2-buten-1-ol	0.1; 3.2	Fruity, green, lavender.	[84, 85]
20	3-Methyl-3-buten-1-ol	0.01–0.1	-	[93]
21	3-Methyl-3-buten-2-ol	0.01–0.1	Herbal, earthy, oily.	[93]
22	3-Methylbutan-1-ol	<0.025; 0.05–0.5	Fusel, oil, alcoholic, whiskey, fruity, banana.	[84, 93]
23	Cuminol (4-Isopropylbenzyl alcohol)	0.032–0.074; 0.023–0.064	Strong, cumin, caraway, spicy, herbal, leather.	[87, 88]
24	Benzyl alcohol	<0.005; 3.5	Floral, rose, phenolic, balsamic.	[69, 84, 85, 86, 91]
25	Ethanol	0.01–0.1; 0.55	Pungent.	[84, 85, 93]
Phenols				
26	4-Vinylphenol	0.3	Chemical, phenolic, medicinal, sweet.	[85]
Aldehydes				
27	(E)-2-Hexenal	0–0.0077; 0.289–0.425	Green, banana, aldehydic, fatty, cheesy.	[69, 83, 85, 86, 87, 88]
28	(Z)-3-Hexenal	2; 0.005–0.05	Green, fatty, grassy, weedy, fruity, apple.	[81, 86]
29	2-Pentenal	0.01–0.1	Pungent, green, fruity, apple, orange, tomato	[83]
30	3-Methyl-2-butenal	0.01–0.1	Sweet, fruity, pungent, brown, nutty, almond, cherry	[83]
31	Hydroxymethylfurfural (5-(Hydroxymethyl)-2-furaldehyde)	0.00143–0.00231	Fatty, buttery, musty, waxy, caramellic.	[69]
32	Acetaldehyde (ethanal)	26	Pungent, ethereal, aldehydic, fruity.	[83]
33	Benzaldehyde	0.2; 0.00165–0.00246	Almond, fruity, powdery, nutty and benzaldehyde-like	[69, 85]
34	Decanal	0–0.00068	Sweet, aldehydic, waxy, orange peel, citrus, floral.	[69]
35	Hexanal	0.005–0.05; 0.1–1	Fresh, green, fatty, aldehydic, grass, leafy, fruity, sweaty.	[69, 83, 85, 86, 87, 88]
Ketones				
36	Diacetyl (2,3-butanedione)	0.05; <0.01	Strong, butter, sweet, creamy, pungent, caramel.	[83, 84]
37	2-Heptanone	0.061–0.102; 0.063–0.108	Fruity, spicy, sweet, herbal, coconut, woody.	
38	2-Nonanone	0.005–0.05; 0.020–0.036	Fresh, sweet, green, weedy, earthy, herbal.	[86, 87, 88]

Table 4. (Continued)

Compounds		Quantitative data (minimum and maximum value mg/kg)	Olfactory descriptor TGSC Information System [93]	References
39	6-Methyl-5-hepten-2-one	3.2	Citrus, green, musty, lemongrass, apple.	[85]
40	3-Hydroxybutanone (acetoin)	0.1–1; 3.6; 0.125–0.749; 0.15; 0.005–0.05	Sweet, buttery, creamy, dairy, milky, fatty.	[69, 83, 84, 85, 86]
41	Propan-2-one (acetone)	0.01–0.1	Solvent, ethereal, apple, pear.	[83]
42	Acetophenone	0.00037–0.00192; 0.005–0.05	Weet, pungent, hawthorn, mimosa, almond, acacia, chemical.	[69, 86]
Lactones				
43	2-Hexen-4-olide (5-ethyl-5 H-furan-2-one)	0.005–0.05	Sweet-fruity, maple, caramel.	[86]
44	γ-Hexalactone (4-hydroxyhexanoic acid lactone)	0.005–0.05; 0.7; 0.05	Herbal, coconut, sweet, coumarin, tobacco	[84, 85, 86]
45	γ-Octalactone (4-hydroxyoctanoic acid lactone)	<0.005; 0.4; 0.05	Sweet, coconut, waxy, creamy, dairy and fatty.	[84, 85, 86]
46	δ-Decalactone	0.005–0.05; 0.476–0.625	Sweet, creamy, lactonic, fatly, coconut, milky	[69, 85, 86, 87, 88]
47	δ-Dodecalactone	0.2	Fatty, sweet, creamy, fruity, peach, apricot, milky and buttery.	[85]
48	δ-Hexalactone	0.6	Creamy, fruity, coconut	[85]
49	δ-Octalactone	0.7	Sweet, coconut, creamy, coumarin and lactonic with a green and fatty nuance.	[85, 87, 88]
Furans				
50	Strawberry ketone (2,5-Dimethyl-4-hydroxy-3(2 H) furanone)	0.1	Weet, slightly burnt brown caramellic, cotton candy with a savory nuance.	[85]
Esters				
51	(Z)-3-Hexenyl acetate	0.004–0.01; 0.003–0.011	Fresh, green, sweet, fruity, with apple, pear and melon nuances	[85, 87]
52	Ethyl 5-hydroxydecanoate	0.8	Sweet, fatty, peach.	[85] (Honkanen et al. 1980)
53	Ethyl 5-hydroxyoctanoate	1.3	Sweet, fatty, pineapple, fruity.	[85]

Compounds		Quantitative data (minimum and maximum value mg/kg)	Olfactory descriptor TGSC Information System [93]	References
54	Ethyl hexanoate	0.005–0.013; 0.005–0.011	Sweet, fruity, pineapple, waxy, fatty a green banana nuance	[87, 88]
55	Methyl nonanoate	0–0.001; 0–0.001	Sweet, fruity, pear-like, waxy, winey, with a slight tropical nuance	[87, 88]
56	2-Methylnaphthalene	0.005–0.05	Sweet, floral, woody.	[86]
57	Acenaphthene	<0.005	Sandalwood, clean, sweet, woody.	[86]
Hydrocarbons				
58	Naphthalene	<0.005	Pungent, dry, tarry.	[86]
Monoterpenes				
59	(Z)-Sabinol	0.2	-	[85]
60	Terpinen-4-ol	0.05–0.5; 0–0.00644	Woody, ceding, mentholic, citrus terpy, spicy.	[69, 85, 86, 87, 88]
61	Camphor	0.005–0.05	Camphoraceous	[86]
62	Geranial	0.005–0.05	Lemon peel, citrus, juicy, green, lime, woody and herbal	[86]
63	Geraniol	0.00209–0.00778; 0.16–1.93	Floral, sweet, rosy, fruity and citronella-like with a citrus nuance	[81, 83, 84, 85, 86, 87, 88, 91]
64	Limonene	0.001–0.002; 0.002	Sweet and citrus peels.	[85, 87, 88]
65	Linalool	0.00124–0.01126; 0.01–0.92	Citrus, orange, floral, terpy, waxy and rose	[69, 84, 85, 86, 87, 88, 91]
66	Myrtenol	0.00007–0.00133	Camphoraceous, woody, cooling, minty with a medicinal nuance	[69]
67	Neral	<0.005	Sweet, citral, lemon peel.	[86]
68	Nerol	0.005–0.05; 0.5; 0.019–0.037; 0.015–0.027	Fresh, citrus, floral, green, sweet, lemon/lime and waxy with a spicy depth.	[85, 86, 87, 88]
69	p-Cymene (1-isopropyl-4-methylbenzene)	0.9; 0.00011–0.00034; 0.008–0.023; 0.012–0.024	Woody, smoky, burnt.	[69, 85, 87]
70	Piperitone	0.005–0.05; 0.2	Herbal, minty, camphor, medicinal.	[85, 86]
71	p-Menthen-2-ol	0.005–0.05	-	[86]
72	Raspberry ketone (4-(4-Hydroxyphenyl)butan-2-one)	3.1; 1.09–4.20	Berry, sweet, woody and raspberry with a ripe, jammy, seedy character.	[85, 91]
73	Sabinene (thuj-4(10)-ene)	0.013–0.032; 0.015–0.030	Woody, spicy, citrus and terpy with green, oily and camphoreous nuances.	[85, 87, 88]
74	Terpinolene	0.7; 0.001–0.004	Sweet, fresh, piney citrus with a woody old lemon peel nuance.	[85, 87]

Table 4. (Continued)

Compounds		Quantitative data (minimum and maximum value mg/kg)	Olfactory descriptor TGSC Information System [93]	References
75	Zingerone	0.3; 0.059–0.234	Spicy, phenolic and balsamic with a creamy vanilla-like nuance.	[85, 87]
76	α-Phellandrene	0–0.0002; 0.026–0.100; 0.028–0.057	Citrus, terpenic, slightly green, black pepper-like.	[69, 85, 87, 88]
77	α-Pinene	0.011–0.027; 0.025–0.033	Woody, piney and turpentine-like, with a slight cooling camphoraceous nuance and a fresh herbal lift.	[85, 87, 88]
78	α-Terpinene	0.011–0.027; 0.004–0.025	Citrusy, woody, terpy with camphoraceous and thymol notes. It has a spicy and juicy citrus nuance.	[87, 88]
79	α-Terpineol	0.012–0.022; 0.035–0.058	Pine-like, woody and resinous with a slight cooling lemon and lime citrus nuance, and a floral dry out.	[69, 85, 87, 88]
80	β-Myrcene	0.001–0.008; 0.004–0.006	Terpy, herbaceous, woody with a rosy celery and carrot nuance.	[87, 88]
81	β-Pinene	0–0.0005	Cooling, woody, piney, and turpentine-like with a fresh minty, eucalyptus and camphoraceous note with a spicy peppery and nutmeg nuance.	[69]
82	γ-Terpinene	0–0.00016; 0.008–0.025	Terpy, sweet, citrus, with tropical and lime nuances	[69, 87, 88]
Sesquiterpenes				
83	(E)-β-Caryophyllene	0.15; 1.2	Sweet, woody, spice, clove, dry.	[84, 85]
84	Caryophyllene oxide	0–0.00933	Sweet, fresh, dry, woody, spicy.	[69]
85	Dihydro-β-ionone	0.1; <0.005	Woody cedar, berry seedy, oily, raspberry with floral, tea and leafy nuances.	[84, 86]
86	Epoxy-β-ionone	0.005–0.05	Fruity, sweet, berry, woody, violet, orris, powdery.	[86]
87	Theaspirane A	0.00006–0.00035	Woody, cooling, minty and camphoreous, with a fresh green herbal nuance.	[69]
88	Theaspirane B	0–0.00037	-	[69]
89	α-Ionol	0.00093–0.00811	Berry, sweet ionone-like, with a woody floral powdery note.	[69]
90	α-Ionone	0.00869–0.01848; 0.1–1	Sweet, woody, floral, violet, orris, tropical, fruity.	[69, 81, 84, 85, 86, 87, 88, 91]

Compounds		Quantitative data (minimum and maximum value mg/kg)	Olfactory descriptor TGSC Information System [93]	References
91	β-Damascenone	<0.005	Woody, sweet, fruity, earthy with green floral nuances.	[86]
92	β-Ionone	0.00858–0.03146; 0.55–2.32	Woody, sweet, fruity, berry-like with a green berry background.	[69, 81, 84, 85, 86, 87, 88, 91]
93	β-Ionol	0.00018–0.00595	Sweet, woody, berry, powdery, floral and fruity.	[69]

Lactones, with a peculiar smell of maple, caramel, coconut, coumarin, and tobacco (among others), are present in small amounts. It was the work of Winter and Enggist [86], that first quantified the lactones 2-hexen-4-olide (5-ethyl-5 H-furan-2-one); γ-hexalactone (4-hydroxyhexanoic acid lactone); γ-octalactone (4-hydroxyoctanoic acid lactone) and the δ-decalactone.

Esters are recognized as having a pleasant fruity smell. Fresh, green, sweet, fruity with apple, pear, peach, pineapple, green banana, and melon nuances. Esthers are truly important for the recognizable fruit aroma and very important for commercial purposes. A few esters were also quantified, being the ethyl 5-hydroxyoctanoate the one that was present in higher concentrations [85].

Monoterpenes are the key contributors to the raspberry sensory characteristics. Their scent is associated with various flowers, fruits, herbs, and spices. Short-chain terpenes (C_{10}), in combination with sesquiterpenes (C_{15}) and diterpenes (C_{20}), display numerous ecological characters; allelopathic, antimicrobial, herbivore deterrent, and pollinator attractants [81, 89].

The first monoterpene quantified in raspberries was the geraniol, by Winter and Sundt [83], and one of the last ones was the myrtenol by Vrhovsek and co-workers in 2014 [69]. One important terpene, important enough to have the raspberry name, is the 4-(4-hydroxyphenyl) butan-2-one (also known as the *raspberry ketone*). Quantified in 1980 and 1991 by Honkanen and co-workers [85, 90] and Larsen and collaborators [91], respectively.

β-Ionone is also an important aromatic compound, found in raspberries in concentrations of 0.00858 to 2.32 mg/kg [69, 81, 84, 86, 87, 88, 90, 91] and that presents pleasant sensory characteristics (woody, sweet, fruity, berry-like with a green berry background).

One interesting and surprising aroma compound, ethyl formate, which presents the characteristic smell of rum and the raspberries flavor, has not been quantified or presented in the literature, has to be an authentic raspberry aroma compound. This ester is formed when ethanol reacts with formic acid. In 2009, researchers at the Cologne University and at the Cornell University detected ethyl formate in outer space, more specifically, in the Sagittarius B2 dust cloud at the center of the Milky Way. Later, scientists in France and Spain confirmed its existence, this time in the Orion constellation [92]. Due to this, it is strange and surprising that the universe "tastes/smells like a raspberry"!

Raspberries Sensory Quality

Consumer demand for organic farming foods is increasing. The preference for organic farming food also reflects an increased interest in personal health, animal well-being, and environmental protection, once the nonexistence of contaminants in the production process. Berry fruits, namely raspberries, have been encouraged for their many health benefits such as antioxidant activity and capacity for lowering the risk of heart diseases and of certain types of cancers. Raspberries are high in vitamin C, vitamin E and, other antioxidants [36].

As referred by several authors [95, 96, 97], production of raspberries, and other berries, is important from the aspect of agricultural development. However, as the times goes, new cultivars have been introduced and agronomic practices have been changed. All these changes, together with factors such as climate, season, soil quality, and degree of ripeness affect the chemical and sensory properties of the berries [98, 99].

Color, Aroma, Flavor and Mouth-Feel Properties

As far as our knowledge is concerned, there aren't any studies published about raspberries produced in organic farming systems and their sensory quality, but, a few have been published about raspberries sensory properties: (i) when aiming to study different package consumer-sized systems [100]; (ii) aiming to study the consumer preference for frozen or fresh red-raspberries using preference mapping [101]; or indirectly, (iii) by determination of sensory properties of red-raspberry extract in fruity flavored beverages comparison to the synthetic preservatives [102]; (iv) by determining the hydrocolloids blend viscosity and sensory properties of raspberry juice-milk [103]; (v) or by studying the acceptance of raspberry-flavored gelatin [104].

When considering raspberry fruits in general, bond to the receptacle plug and raspberry color are the main criteria used by the producers for assessment of the right time for harvesting, whereas color is also the main criterion used by the consumer for judging raspberry quality [105]. Color charts can be used to quantify the visible color, Figure 2 [104].

In the work of Krüger et al. [106] – "Effects of ripening stage and storage time on quality parameters of red-raspberry fruit," sensory descriptive analysis showed an effect of progressed ripening on odor and taste, moreover, acceptance of raspberry-fruits was strongly correlated to the sensory attributes fruity-taste, sweetness, and sweetness: acidity ratio.

But, if in raspberries no studies have been made about the sensory properties imprinted by organic farming, in other berries, like strawberries, and blackberries, some works have been published.

Organic and conventional farming of strawberries were grown in Spain, in head-to-head plots, under identical environmental conditions [107]. The organically grown berries had superior quality when compared to the conventional ones, showing an intense color, higher sugar, and better sensory properties. In another study, in California, organic farming strawberries were sweeter and preferred by consumers compared to conventionally grown berries [108].

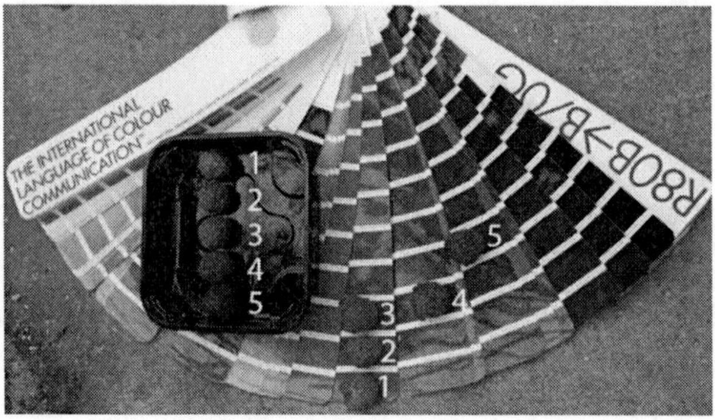

Figure 2. Numbered colors used for maturity stage classification, at harvest, in ‘Glen Ample’ raspberry variety, using a color chart [104].

Reganold et al. [109], to determine if differences in strawberries quality exist when organic and conventional farming strawberries are compared, analyzed 42 fruit, 11 leaf, and 6 sensory properties several times for two years of study. For each field of study, a strawberry pair (organic and conventional farming) was analyzed simultaneously, at the same stage of harvest maturity, under equal storage conditions and transportation methods. Using hedonic/intensity ratings, the consumer-sensory panel, in this study, found organic farming strawberries to be sweeter and have preferable flavor, appearance, and overall acceptance compared to conventional farming strawberries.

Nevertheless, and according to Kårlund and co-workers opinion [110], the results indicate that the chemical composition and sensory properties of strawberries grown in different farming systems might vary according to cultivar. The same can be extended for blackberries [12]. After having studied the influence of cultivar, conventional and organic farming practices on the phenolic and sensory profile of blackberries (*Rubus fruticosus*) it can be concluded that the effect of farming practices on the blackberries’ chemical and sensory profile dependend on the variety and could not be generalized.

ORGANIC FARMING RASPBERRIES CONSUMERS' PERCEPTION

Measuring food sensory quality is tricky and intrinsically subjective. Research indicates that when consumers read "organic farming" in the food product label, the product acceptability increases [111]. This phenomenon is known as the *halo effect* or *cognitive bias* [112]. Results from sensory evaluation tests can vary: where there used trained or untrained panelists? Were the food samples handled in the same way? These questions must be addressed before the tasting sessions.

The "Comparative report from qualitative consumer research" by Stolz et al. [113] relates that consumers were requested to describe their sensory experiences when consuming organic and conventional farming food products. The parameters "taste," "texture" and "appearance" were very often described whereas the category "odor" was less often cited. Moreover, most of the descriptions mentioned were positive experiences with organic farming products. However, *light users* of organic farming products tended to mention more often negative quality parameters (e.g., shorter shelf-life or visible quality matters; or negative perceived sensory attributes).

The same report [112] mentioned that the predominant sensory expectation of organic products is that their taste is more "authentic." Being "authentic" the taste of a product as it used to be instead of the taste of regular products as these are presently sold; or expected products to taste and feel like the products from the farm or from their childhood; or more intense, pure or natural taste; or more natural and less industrial / manufactured products, since conventional products on the market today are perceived rather as tasteless, neutral or bloated [112].

Recently, Laaksonen and co-workers [114], studied the effects of sensory quality and consumer characteristics on consumer acceptance of berry fruits. Consumers completed an online questionnaire scoring the liking, usage, and familiarity of different berries and then the Food Neophobia – FNS [115] and General Health Interest (GHI) scales [116].

Participants completed a sensory test where they profiled berries using Projective Mapping. At the end of the study it was verified that berries were classified according to liking ratings and sensory properties; *mostly liked* berries were 'sweet' and not 'sour' or 'bitter'; *less-liked* berries were either 'sour' and 'bitter' or "unfamiliar" [113]. In the same work was also found that ladies and older participants gave higher liking ratings for berries, and food neophobia only affected the acceptance of generally *less-liked* berries.

CONCLUSION

Sensory quality is a combination of different senses of perception felt when choosing and eating food. Today, consumers of organic farming food tend to believe that it tastes better than its counterpart. Research indicates that consumers view food that is organically farmed to taste better (more sweet, less sour or bitter) and have more flavors compared to its non-organic alternatives.

The flavor of berries has not been studied as well as other fruits. Although the environment alters the flavor quality, genetic factors determine the flavor profile quantitatively and qualitatively.

Future research on berry quality must adopt an interdisciplinary approach so that the genomic tools related to flavor formation in berries can be developed and utilized and the differences between different farming systems can be properly quantified in terms of chemical and sensory parameters.

REFERENCES

[1] Vilela, A., Pinto, T., Anjos, A., Cosme, F. "Chemical and sensorial characteristics of red fruits imprinted by organic farming production" in *Agricultural Research Updates*, Vol 19, (2017). eds

P. Gorawala and S. Mandhatri, Nova Science Publishers, Inc. ISBN: 978-1-53611-012-8.

[2] Mie, A., Andersen, H. A., Gunnarsson, S., Kahl, J., Kesse-Guyot, E., Rembiałkowska, E., Quaglio, G., Grandjean, P. "Human health implications of organic food and organic agriculture: a comprehensive review." *Environ Health* 16, (2017): 111. doi:10. 1186/s12940-017-0315-4.

[3] Benbrook, C. "The impacts of yield on nutritional quality: lessons from organic farming." *Hortscience* 44, (2009): 12–14. doi.org/10. 1155/2015/431487.

[4] Burlingame, B., Dernini, S., Nutrition and Consumer Protection Division; FAO (eds.). (2010). Sustainable diets and biodiversity: directions and solutions for policy, research and action. *Proceedings of the International Scientific Symposium on Biodiversity and Sustainable Diets: United Against Hunger; 2010 Nov 3–5; Rome.* Rome: Food and Agriculture Organization. E-ISBN 978-92-5-107288-2 (PDF).

[5] WHO (World Health Organization) (2012). Health indicators of sustainable agriculture, food and nutrition security in the context of the Rio+20, *UN Conference on Sustainable Development*. www.who.int/hia/green_economy/indicators_food.pdf. 2012 (Accessed 13 November, 2018).

[6] Vogtmann, H. (1991). In Lassen J. Paper – *Food Quality and the Consumers*. 1993.

[7] Luning, P. A., Marcelis, W. J., Jongen, W. M. (2005). "Zarządzanie jakością żywności. Ujęcie technologiczno-menadżerskie." ["Food quality management. Technological and managerial approach."] *Wydawnictwo Naukowo-Techniczne, Warszawa*. ISBN: 83-204-2966-8.

[8] FAO/FAOSTAT (2015): *Food and Agriculture Organization of the United Nations FAOSTAT Online Database* (Accessed 01 August, 2015); http://faostat.fao.org/site/377/default.aspx#ancor.

[9] Średnicka-Tober, D., Barański, M., Seal, C., Sanderson, R., Benbrook, C., Steinshamn, H., Leifert, C. "Composition differences

between organic and conventional meat: A systematic literature review and meta-analysis." *British Journal of Nutrition* 115(6), (2016): 994-1011. doi:10.1017/S0007114515005073.

[10] Willer, H., Schaack, D., Lernoud, J. (2017). "Organic farming and market development in Europe and the European Union" in the *World of Organic Agriculture - Statistics and Emerging Trends 2017,* eds H. Willer; J. Lernoud; Frick and Bonn: FiBL and IFOAM. https://shop.fibl.org/CHen/mwdownloads/download/link/id/785/?ref =1.

[11] *Eurostat* (2018). http://ec.europa.eu/eurostat. (Accessed 16 November, 2018).

[12] Pinto, T., Vilela, A., Pinto, A., Nunes, F. M., Cosme, F., Anjos, R. "Influence of cultivar and of conventional and organic agricultural practices on phenolic and sensory profile of blackberries (*Rubus fruticosus*)." *Journal of the Science of Food and Agriculture* 98, (2018): 4616-4624. doi:10.1002/jsfa.8990.

[13] Acs, S., Berentsen, P. B. M., Huirne, R. B. M. "Modelling conventional and organic farming: a literature review." *NJAS - Wageningen Journal of Life Sciences* 53 (1), (2005): 1–18.

[14] Morison, J. I. L., Baker, N. R., Mullineaux, P. M., Davies, W. J. "Improving water use in crop production." *Philos Trans B*, 363, (2008): 639–658. doi:10.1098/rstb.2007.2175.

[15] IFOAM (International Federation of Organic Agriculture Movements) (2014). *The IFOAM Norms for Organic Production and Processing, Version 2014*. Germany. http://www.ifoam.bio/sites/ default/files/ifoam_norms_july_2014_t.pdf.

[16] Govindaraj, M., Vetriventhan, M., Srinivasan, M. "Importance of genetic diversity assessment in crop plants and its recent advances: An overview of its analytical perspectives." *Genetics Research International* (2015):1-14. doi: 10.1155/2015/431487.

[17] Ceglie, F. G., Amodio, M. L., Colelli, G. "Effect of organic production systems on quality and postharvest performance of horticultural produce." *Horticulturae*, 2(2), (2016): 4; https://doi.org/ 10.3390/horticulturae2020004.

[18] Winter, C. K., Davis, S. F. Organic foods. *Journal of Food Science* 71(9), (2006): R117–R124.

[19] Bourn, D., Prescott, J. "A comparison of the nutritional value, sensory qualities, and food safety of organically and conventionally produced foods." *Critical Reviews in Food Science and Nutrition* 42 (1), (2002): 1–34.

[20] Fuller, R. J., Norton, L. R., Feber, R. E., Johnson, P. J., Chamberlain, D. E., Joys, A. C., Mathews, F., Stuart, R. C., Townsend, M. C., Manley, W. J., Wolfe, M. S., Macdonald, D. W., Firbank, L. G. "Benefits of organic farming to biodiversity vary among taxa." *Biology Letters 1*, (2005): 431–434.

[21] Hargreaves, J., Adl, M. S., Warman, P. R., Rupasinghe, H. P. V. "The effects of organic amendments on mineral element uptake and fruit quality of raspberries." *Plant Soil* 308, (2008): 213–226.

[22] Rao, A. V., Snyder, D. M. "Raspberries and human health: A review." *Journal of Agricultural and Food Chemistry* 58, (2010): 3871–3883.

[23] Oliveira, P. B., Lopes-Da-Fonseca, L., Monteiro, A. A. "Combining different growing techniques for all year round red raspberry production in Portugal." *Acta Horticulturae* 585, (2002): 545-553.

[24] Dale, A; Sample, A., King, E. "Breaking dormancy in red raspberries for greenhouse production." *Hortscience* 38, (2003): 515-519.

[25] Dale, A., Pirgozliev, S., King, E. M., Sample, A. "Scheduling primocane-fruiting raspberries (*Rubus idaeus* L.) for year-round production in greenhouses by chilling and Summer-pruning of canes." *Journal of Horticultural Science & Biotechnology* 80, (2005): 346-350.

[26] Gotame, T. P. (2014). *Understanding the effects of temperature on raspberry physiology and gene expression profiles*. PhD Thesis, Faculty of Science and Technology, Department of Food Science, Aarhus University, Denmark. https://pure.au.dk/ws/files/69019160/ Understanding_the_effects_of_temperature_on_raspberry_physiolog y_and_gene_TGO.pdf.

[27] Ballington, J. R., Fernandez, G. E. "Breeding raspberries adapted to warm humid climate with fluctuating temperature in winter." *Acta Horticulturae* 777, (2008): 87-90.

[28] Marlett, J. A., Vollendorf, N. W. "Dietary fibre content and composition of different forms of fruits." *Food Chemistry* 51, (1994): 39–44.

[29] Schmeda-Hirschmann, G., Feresin, G., Tapia, A., Hilgert, N., Theoduloz, C. "Proximate composition and free radical scavenging activity of edible fruits from the Argentinian Yungas." *Journal of the Science of Food and Agriculture*, 85, (2005): 1357–1364.

[30] Parry, J., Su, L., Luther, M., Zhou, K., Yurawecz, M. P., Whittaker, P., Yu L. "Fatty acid composition and antioxidant properties of cold-pressed marionberry, boysenberry, red raspberry, and blueberry seed oils." *Journal of Agricultural and Food Chemistry* 53, (2005): 566–573.

[31] Beekwilder, J., Jonker, H., Meesters, P., Hall, R. D., Van Der Meer, I. M., De Vos, C. H. R. "Antioxidants in raspberry: on-line analysis links antioxidant activity to a diversity of individual metabolites." *Journal of Agriculture and Food Chemistry* 53, (2005): 3, 313–320.

[32] Król, K., Orzel, A., Jagla, J. "Evaluation of twenty raspberry and blackberry cultivars grown in thesouth of Poland." ["Evaluation of twenty raspberry and blackberry cultivars grown in thesouth of Poland."]. *Zeszyty Naukowe Instytutu Sadownictwa i Kwiaciarstwa* 16, (2008): 125-34.

[33] Czech, A., Rusinek, E., Merska, M. "Content of selected bio-elements in berry fruit and juices." *Problemy Higieny i Epidemiologii* 92(4), (2011): 836-839. ["Content of selected bio-elements in berry fruit and juices." *Problems of Hygiene and Epidemiology* 92(4), (2011): 836-839].

[34] Lee, J., Dossett, M., Finn, C. E. "Rubus fruit phenolic research: The good, the bad, and the confusing." *Food Chemistry* 130 (2012): 785-796.

[35] Kalt, W., Forney, C. F., Martin, A., Prior, R. L. "Antioxidant capacity, vitamin C, phenolics, and anthocyanins after storage of

small fruits." *Journal of Agricultural and Food Chemistry* 47, (1999): 4638–4644.

[36] Kazimierczak, R., Hallmann, E., Kowalska, K., Rembiałkowska, E. "Biocompounds content in organic and conventional raspberry fruits." *Acta Fytotechnica et Zootechnica* 18 (Special Issue), (2015): 40–42. doi:10.15414/afz.2015.18.si.40-42. ["Biocompounds content in organic and conventional raspberry fruits." *Scientific Journal for phytotechnics and zootechnics* 18 (Special Issue), (2015): 40–42. doi:10.15414/afz.2015.18.si.40-42].

[37] Zhang, Z., Zhu, X., Xie, Z., Zhao, Z., Dubé, C., Roussel, D., Charles, M. T., Khanizadeh, S. "Yield, fruit quality, antioxidant capacity and phenolic composition of advancedraspberry lines from Quebec." *Journal of Food, Agriculture & Environment* 11(2), (2013): 278-284.

[38] Ciesliks, E., Greda, A. Adamus, W. "Contents of polyphenols in fruit and vegetables." *Food Chemistry* 94, (2006); 135–142.

[39] Zafra-Stone, S., Yasmin, T., Bagchi, M., Chatterjee, A., Vinson, J.A., Bagchi, D. "Berry anthocyanins as novel antioxidants in human health and disease prevention." *Molecular Nutrition & Food Research* 51(6), (2007): 675-83.

[40] Tal, A. "Making conventional agriculture environmentally friendly: Moving beyond the glorification of organic agriculture and the demonization of conventional agriculture." *Sustainability*, 10, (2018): 1078. https://doi:10.3390/su10041078.

[41] Bobinaite, R., Viskelis, P., Venskutonis, P. R. "Variation of total phenolics, anthocyanins, ellagic acid and radical scavenging capacity in various raspberry (*Rubus* spp.) cultivars." *Food Chemistry* 132(3), (2012): 1495-1501.

[42] Skupien, K., Ochmian, I., Grajkowski, J., Krzywy-Gawronska, E. "Nutrients, antioxidants, and antioxidant activity of organically and conventionally grown raspberries." *Journal of Applied Botany and Food Quality-Angewandte Botanik* 84(1), (2011): 85-89.

[43] Pantelidisa, G. E., Vasilakakisa, M., Manganaris, G. A., "Diamantidisa, G. Antioxidant capacity, phenol, anthocyanin and ascorbic acid contents in raspberries, blackberries, red currants,

gooseberries and Cornelian cherries." *Food Chemistry* 102(3), (2007): 777-783.

[44] Vool, E., Karp, K., Moor, U., Starast, M. "Yield quality in some taxona of the genus *Rubus* depending on the cultivation technology." *European Journal of Horticultural Science* 72(1), (2007): 32-38.

[45] Remberg, S. F., Sonsteby, A., Aaby, K., Heide, O. M. "Influence of postflowering temperature on fruit size and chemical composition of Glen Ample Raspberry (*Rubus idaeus* L.)." *Journal of Agricultural and Food Chemistr*y 58(16), (2010): 9120-9128.

[46] Ancos, B. d, Gonzàlez, E. M., Cano, P. "Ellagic acid, vitamin C, and total phenolic contents and radical scavenging capacity affected by freezing and frozen storage in raspberry fruit." *Journal of Agricultural and Food Chemistry* 48, (2000): 4565–4570.

[47] Kumar, D., Rizvi, S. I. Significance of vitamin C in human health and disease. *Annals of Phytomedicine* 1, (2012): 9-13.

[48] Borges, G., Degeneve, A., Mullen, W., Crozier, A. "Identification of flavonoid and phenolic antioxidants in black currants, blueberries, raspberries, red currants, and cranberries." *Journal of Agriculture and Food Chemistry* 58, (2010): 3901-3909.

[49] Mullen, W., McGinn, J., Lean, M. E., MacLean, M. R., Gardner, P., Duthie, G. G., Yokota, T., Crozier, A. Ellagitannins, flavonoids, and other phenolics in red raspberries and their contribution to antioxidant capacity and vasorelaxation properties. *Journal of Agricultural and Food Chemistry* 50 (2002): 5191-5196.

[50] Plessi, M., Bertelli, D., Albasini, A. "Distribution of metals and phenolic compounds as a criterion to evaluate variety of berries and related jams." *Food Chemistry*, 100, (2007): 419–427.

[51] Cekic, C., Ozgen, M. "Comparison of antioxidant capacity and phytochemical properties of wild and cultivated red raspberries (*Rubus idaeus* L.)." *Journal of Food Composition and Analysis* 23(6), (2010): 540-544.

[52] Milivojevic, J. M., Nikolic, M. D., Maksimovic, J. J. D., Radivojevic, D. D. "Generative and fruit quality characteristics of

primocane fruiting red raspberry cultivars." *Turkish Journal of Agriculture and Forestry* 35(3), (2011): 289-296.

[53] Herrmann, K. "Über die Gehalte der hauptsächlichen Pflanzenphenole im Obst." *Flüssiges Obst* 59, 1992: 66–70. ["About the levels of the main vegetable phenols in the fruit. " *Liquid fruit* 59, 1992: 66–70].

[54] Heinonen, I. M., Meyer, A. S., Frankel, E. N. "Antioxidant activity of berry phenolics on human low-density lipoprotein and liposome oxidation." *Journal of the Science of Food and Agriculture* 46, (1998): 4107–4112.

[55] Daniel, E. M., Krubnick, A. S., Heur, Y. H., Blinzler, J. A., Nims, R. W., Stoner, G. D. "Extraction, stability, and quantitation of ellagic acid in various fruits and nuts." *Journal of Food Composition and Analysis* 2, (1989): 338–349.

[56] Hollman, P. C. H., Venema, D. P. "The content of the potentially anticarcinogenic ellagic acid in plant foods." In: Waldron KW, Johnson IT, Fenwick GR, eds. *Food and Cancer Prevention: Chemical and Biological Aspects*. Cambridge, UK: Royal Society of Chemistry, (1993), p. 203–208.

[57] Starke, H., Herrmann, K. "The phenolics of fruits. VIII. Changes in flavonol concentrations during fruit development." *Zeitschrift für Lebensmittel-Untersuchung und -Forschung* 161, (1976): 131–135. ["The phenolics of fruits. VIII. Changes in flavonol concentrations during fruit development." *Journal of Food Examination and Food Research*, 161, (1976): 131–135].

[58] Justesen, U., Knuthsen, P., Leth, T. "Quantitative analysis of flavonols, flavones, and flavanones in fruits, vegetables and beverages by high-performance liquid chromatography with photo-diode array and mass spectrometric detection." *Journal of Chromatography A* 799, (1998): 101–110.

[59] Torre, L. C., Barrit, B. H. "Quantitative evaluation of Rubus fruit anthocyanin pigments. " *Journal of Food Science* 42 (1977): 488–490.

[60] Wang, S. Y., Lin, H. S. "Antioxidant activity in fruits and leaves of blackberry, raspberry, and strawberry varies with cultivar and developmental stage *Journal of Agricultural and Food Chemistry* 48, (2000):140–146.

[61] Arts, I. C., van de Putte, B., Hollman, P. C. H. "Catechin contents of foods commonly consumed in the Netherlands. 1. Fruits, vegetables, stable foods, and processed foods. "*Journal of Agricultural and Food Chemistry* 48, (2000): 1746–1751.

[62] Schuster, B., Herrmann, K. "Hydroxybenzoic and hydroxycinnamic acid derivatives in soft fruits." *Phytochemistry* 24, (1985): 2761–2764.

[63] Mosel, H. D., Herrmann, K. "Phenolics of fruits. IV. The phenolics of blackberries and raspberries and their changes during development and ripeness of the fruits." *Zeitschrift für Lebensmittel-Untersuchung und -Forschung* 1974; 154: 324–327. ["Phenolics of fruits. IV. The phenolics of blackberries and raspberries and their changes during development and ripeness of the fruits." *Journal of Food Examination and Food Research* 1974; 154: 324–327].

[64] Haslam, E. (1998). *Practical Polyphenolics. From Structure to Molecular. Recognition and Physiological Action*. Cambridge University Press: Cambridge, UK.

[65] Gasperotti, M., Masuero, D., Vrhovsek, U., Guella, G., Mattivi, F. Profiling and accurate quantification of Rubus ellagitannins and ellagic acid conjugates using direct UPLC-Q-TOF HDMS and HPLC-DAD analysis. *Journal of Agricultural and Food Chemistry*, 58, (2010): 4602–4616.

[66] Mullen, W., Yokota, T., Lean, M. E, Crozier, A. "Analysis of ellagitannins and conjugates of ellagic acid and quercetin in raspberry fruits by LC-MSn." *Phytochemistry* 64, (2003): 617-624.

[67] Hakkinen S. H., Karenlampi S. O., Heinonen I. M., Mykkanen H. M., Torronen A. R. "Screening of selected flavonoids and phenolic acids in 19 berries." *Food Research International*, 32, (1999): 345–353.

[68] Jakobek, L., Seruga, M., Seruga, B., Novak, I., Medvidovic-Kosanovic, M. Phenolic compound composition and antioxidant activity of fruits of Rubus and Prunus species from Croatia. *International Journal of Food Science and Technology*, 44, (2009): 860–868.

[69] Vrhovsek, U., Lotti, C., Masuero, D., Carlin, S., Weingart, G., Mattivi, F. "Quantitative metabolic profiling of grape, apple and raspberry volatile compounds (VOCs) using a GC/MS/MS method." *Journal of Chromatography B* 966, (2014): 132–139.

[70] Priyadarsini, K. J., Khopde, S. M., Kumar, S. S., Mohan, H. "Free radical studies of ellagic acid, a natural phenolic antioxidant. " *Journal of Agriculture and Food Chemistry* 50 (2002): 200–206.

[71] Vattem, D. A., Shetty, K. "Biological functionality of ellagic acid: a review." *Journal of Food Biochemistry* 29, (2005): 234–266.

[72] Scalbert, A., Clifford, N., Michael, A. "Ellagitannins - nature, occurrence and dietary burden. " *Journal of the Science of Food and Agriculture* 80(7), (2000):1118-1125.

[73] Tomas-Barberan, F. A., Clifford, M. N. "Dietary hydroxybenzoic acid derivatives - nature, occurrence and dietary burden. " *Journal of the Science of Food and Agriculture* 80(7), (2000): 1024-1032.

[74] Wildanger, W., Herrmann, K. "The phenolics of fruits. II. The favonols of fruits. " *Zeitschrift für Lebensmittel-Untersuchung und -Forschung*, 151, (1973): 103-108. ["The phenolics of fruits. II. The favonols of fruits. " *Journal of Food Examination and Food Research*, 151, (1973): 103-108].

[75] González, E. M., Ancos, B. d., Cano, M. P. "Relation between bioactive compounds and free radical-scavenging capacity in berry fruits during frozen storage." *Journal of the Science of Food and Agriculture* 83(7), (2003): 722-726.

[76] Maatta-Riihinen, K. R., Kamal-Eldin, A., Torronen, A. R. "Identification and quantification of phenolic compounds in berries of Fragaria and Rubus species (Family Rosaceae)." *Journal of Agricultural and Food Chemistry* 52, (2004): 6178–6187.

[77] Kassim, A., Poette, J., Paterson, A., Zait, D., McCallum, S., Woodhead, M., Smith, K., Hackett, C., Graham, J. "Environmental and seasonal influences on red raspberry anthocyanin antioxidant contents and identification of quantitative traits loci (QTL)." *Molecular Nutrition and Food Research*, 53, (2009): 625–634.

[78] Gu, L., Hammerstone, J. F., Beecher, G., Holden, J., Haytowitz, D., Gebhardt, S., Prior, R. L. "Concentration of proanthocyanidins in common foods and estimates of consumption. " *Journal of Nutrition* 134, (2004): 613–617.

[79] Pawlowska, M. A., Fabiano, C., Alessandra, B. "Qual_quantitative analysis of cornus mus L (Cornaceae) fruits." *Food Chemistry* 115, (2009): 450-455.

[80] Robertson, G. W., Griffiths, D. W., Woodford, J. A.T., Birch, A. N. E. "Changes in the chemical composition of volatiles released by the flowers and fruits of the red raspberry (*Rubus idaeus*) cultivar glen prosen." *Phytochemistry* 38, (1995): 1175–1179.

[81] Aprea, E., Carlin, S., Giongo, L., Grisenti, M., Gasperi, F." Characterization of 14 raspberry cultivars by Solid-Phase Microextraction and relationship with gray mold susceptibility." *Journal of Agricultural and Food Chemistry.* 58, (2010): 1100–1105.

[82] Polashock, J. J., Saftner, R. A., Kramer M. "Postharvest highbush blueberry fruit antimicrobial volatile profiles in relation to anthracnose fruit rot resistance." *Journal of the American Society for Horticultural Science* 132 (6), (2007): 859-868.

[83] Winter, M., Sundt, E. "Recherches sur les arŏmes." 5e communication. Analyse de l'arŏme des framboises. I. [Research on aroma. "*5th Communication, Analysis of the Aroma of Raspberries I*]. Les constituants carbonylés volatils. HCA, 45, (1962): 2195–2211.

[84] Pyysalo, T. "Identification of volatile compounds in hybrids between raspberry (*Rubus idaeus,* L.) and arctic bramble (*Rubus arciticus,* L.)." ["Identification of volatile compounds in hybrids between raspberry (Rubus idaeus, L.) and arctic bramble (Rubus arciticus,

L.).”]. *Zeitschrift für Lebensmittel-Untersuchung und -Forschung* 162, (1976): 263–272.

[85] Honkanen, E., Pyysalo, T., Hirvi, T. “The aroma of finnish wild raspberries, *Rubus idaeus*, L.” *Zeitschrift für Lebensmittel-Untersuchung und -Forschung* 171, (1980): 180–182. [“The aroma of finnish wild raspberries, Rubus idaeus, L.” *Journal of Food Examination and Food Research* 171, (1980): 180–182].

[86] Winter, M., Enggist, P. *“Recherches sur les arǒmes.”* 17e communication. Sur l’arǒme de framboise. [“*Research on aromas*.” 17th communication. On the aroma of raspberry], IV. HCA, 54, (1971): 1891–1898.

[87] Malowicki, S. M. M., Martin, R., Qian, M. C. “Volatile composition in raspberry cultivars grown in the Pacific Northwest determined by Stir Bar Sorptive Extraction–Gas Chromatography–Mass Spectrometry.” *Journal of Agricultural and Food Chemistry* 56, (2008): 4128–4133.

[88] Malowicki, S. M. M., Martin, R., Qian, M. C. “Comparison of sugar, acids, and volatile composition in raspberry bushy dwarf virus-resistant transgenic raspberries and the wild type “Meeker” (*Rubus Idaeus* L.).” *Journal of Agricultural and Food Chemistry* 56, (2008): 6648–6655.

[89] Davis, E. M. “Advances in the enzymology of monoterpene cyclization reactions.” In Compr*ehensive Natural Products II,* Edited by Hung-Wen (Ben) Liu, Lew Mander, Elsevier, 585-608. (2010). https://doi.org/10.1016/B978-008045382-8.00004-6.

[90] Honkanen, E., Hirvi, Y. “The flavor of berries,” in *Food Flavours*, eds I. Morton and A. Macleod (Amsterdam: Elsevier Scientific Publications), (1990), 125–193.

[91] Larsen, M., Poll, L., Callesen, O., Lewis, M. “Relations between the content of aroma compounds and the sensory evaluation of 10 raspberry varieties (*Rubus idaeus* L).” *Acta Agriculturae Scandinavica* 41, (1991): 447–454.

[92] Sample, I. "Galaxy's centre tastes of raspberries and smells of rum, say astronomers." *The Guardian*. (2009). Archived from the original on 6 July 2017.

[93] TGSC Information System (2018). *The Good Scents Company Information System*. http://www.thegoodscentscompany.com.

[94] Sundt, E., Winter, M. "*Recherches sur les arŏmes*" .6e communication. Analyse de l'arŏme volatils des framboises II. Les alcools ["Research on aromes." 6th communication. Analysis of the volatile aroma of raspberries II]. *HCA*, 45, (1962): 2212–2218.

[95] Ercisli, S. "Rose germplasm resources of Turkey." *Genetic Resources and Crop Evolution* 52, (2004): 787-795.

[96] Ercisli, S., Tosun, M. Duralija, B., Voca, S., Sengul, M., Turan, M. "Phytochemical content of some black (*Morus nigra* L.) and purple (*Morus rubra* L.) mulberry genotypes." *Food Technology and Biotechnology*. 48, (2010): 102-106.

[97] Zorenc, Z., Veberic, R., Stampar, F., Koron, D., Mikulic-Petkovsek, M. Changes in berry quality of northern highbush blueberry (*Vaccinium corymbosum* L.) during the harvest season. *Turkish Journal of Agriculture and Forestry* 40, (2016): 855-867.

[98] Pelayo, C., Ebeler, S. E., Kader, A. A. "Postharvest life and flavor quality of three strawberry cultivars kept at 5°C in air or air+20 kPa CO_2." *Postharvest Biology and Technology* 27, (2003): 171-183.

[99] Skrovankova, S., Sumczynski, D., Mlcek, J., Jurikova, T., Sochor, J. "Bioactive compounds and antioxidant activity in different types of berries." *International Journal of Molecular Sciences* 16, (2015): 24673-24706.

[100] Van der Steen, C., Jacxsens, L., Devlieghere, F., Debevere, J. "Combining high oxygen atmospheres with low oxygen modified atmosphere packaging to improve the keeping quality of strawberries and raspberries." *Postharvest Biology and Technology* 26(1), (2002): 49–58. doi:10.1016/s0925-5214(02)00005-4.

[101] Villamor, R. R., Daniels, C. H., Moore, P. P., Ross, C. F. "Preference mapping of frozen and fresh Raspberries." *Journal of*

Food Science, 78(6), (2013): S911–S919. doi:10.1111/1750-3841.12125.

[102] Ozarda, O., Demirkoz, A. B., Özdemir, M. "Sensory characteristics and antioxidant capacity of red raspberry extract as a preservative in fruity flavored beverages." *Journal of Food Science and Technology*, 52(10), (2015): 6687–6694. doi:10.1007/s13197-015-1763-x.

[103] Abedi, F., Sani, A. M., Karazhiyan, H. "Effect of some hydrocolloids blend on viscosity and sensory properties of raspberry juice-milk." *Journal of Food Science and Technology* 51(9), (2012): 2246–2250. doi:10.1007/s13197-012-0705-0.

[104] Palazzo, A. B., Bolini, H. M. A. "Multiple time-intensity analysis and acceptance of raspberry-flavored gelatin." *Journal Sensory Studies* 24(5), (2009): 648–663. doi:10.1111/j.1745-459x.2009.00231.x.

[105] Stavang, J. A., Freitag, S., Foito, A., Verrall, S., Heide, O.M., Stewart, D., Sønsteby, A. "Raspberry fruit quality changes during ripening and storage as assessed by colour, sensory evaluation, and chemical analyses." *Scientia Horticulturae* 195, (2015): 216-225. https://doi.org/10.1016/j.scienta.2015.08.045.

[106] Krüger, E., Schopplein, E., Rasim, S., Cocca, G., Fischer, H. " Effects of ripening stage and storage time on quality parameters of red raspberry fruit." *European Journal of Horticultural Science* 68, (2003): 176–182.

[107] Cayuela, J. A., Vidueira, J. M., Albi, M. A., Gutierrez, F. "Influence of the ecological cultivation of strawberries (*Fragaria* × *Ananassa* Cv. Chandler) on the quality of the fruit and on their capacity for conservation." *Journal of Agriculture and Food Chemistry* 45(5), (1997): 1736-1740.

[108] Kristiansen, P., Taji, A., Reganold J. (eds.). *Organic Agriculture: A Global Perspective.* CSIRO Publishing, Collingwood, Victoria, Australia. (2006). 480 pp.

[109] Reganold, J. P., Andrews, P. K., Reeve, J. R., Carpenter-Boggs, L., Schadt, C. W., Alldredge, J. R., Ross, C. F., Davies, N. M., Zhou, J. "Fruit and soil quality of organic and conventional strawberry

agroecosystems." *PloS one*, 5(9), (2010): e12346. doi:10.1371/journal.pone.0012346.

[110] Kårlund, A., Hanhineva, K., Lehtonen, M., Karjalainen, R. O., Sandell, M. "Nontargeted metabolite profiles and sensory properties of strawberry cultivars grown both organically and conventionally." *Journal of Agricultural and Food Chemistry* 63(3), (2015): 1010-1019. doi: 10.1021/jf505183j.

[111] Moskowitz, H. R., Beckley, J. H., Resurreccion A. V. A. In: *Sensory and Consumer Research in Food Product Design and Development*. Second edition, Blackwell Publishing Ltd. and the Institute of Food Technologists, (2012). https://doi.org/10.1002/9781119945970.

[112] Rosenzweig P. M. *The halo effect and the eight other business delusions that deceive managers*. New York, NY: Free Press. pp. xi. (2014). ISBN 978-1-4767-8403-8.

[113] Stolz, H., Jahrl, I., Baumgart, L., Schneider, F. (2010). Sensory experiences and expectations of organic food. Results of Focus Group Discussions. In *Comparative report from qualitative consumer research Deliverable 4.2*, EU Project ECROPOLIS – Contract No. 218477-2 (FP 7). Research Institute of Organic Agriculture (FiBL), Frick and Frankfurt.

[114] Laaksonen, O., Knaapila, A., Niva, T., Deegan, K. C., Sandell, M. "Sensory properties and consumer characteristics contributing to liking of berries." *Food Quality and Preference* 53, (2016): 117–126. doi:10.1016/j.foodqual.2016.06.004.

[115] Damsbo-Svendse, M., Frøst, M. B., Olsen, A. "A review of instruments developed to measure food neophobia." *Appetite* 113, (2017): 358-367. https://doi.org/10.1016/j.appet.2017.02.032.

[116] Urala, N., Lähteenmäki, L. "Attitudes behind consumers' willingness to use functional foods." *Food Quality and Preference* 15(7), (2004): 793-803. doi: 10.1016/j.foodqual.2004.02.008.

Contents of Earlier Volumes

Agricultural Research Updates: Volume 26

Agricultural Research Updates, Volume 25

Agricultural Research Updates: Volume 24

Agricultural Research Updates: Volume 23

Agricultural Research Updates: Volume 22

INDEX

A

B

C

D

E

I

K

L

M

N

O

P

Q

R

S

T